# Review Questions for
# HUMAN
# PHYSIOLOGY

**Dedication**

To James and Jonathan

# Review Questions for
# HUMAN
# PHYSIOLOGY

**James N. Pasley, PhD**

University of Arkansas College of Medicine

**Review Questions Series**
Series Editor: Thomas R. Gest, PhD
University of Arkansas for Medical Sciences

## The Parthenon Publishing Group
International Publishers in Medicine, Science & Technology

NEW YORK                                    LONDON

Published in the USA by
The Parthenon Publishing Group Inc.
One Blue Hill Plaza,
PO Box 1564, Pearl River,
New York 10965, USA

Published in Europe by
The Parthenon Publishing Group Limited
Casterton Hall, Carnforth,
Lancs LA6 2LA, UK

**Library of Congress Cataloging-in-Publication Data**

Pasley, James N.
    Review questions for human physiology / by James N. Pasley
        p.    cm. -- (Review questions series)
    ISBN: 1-85070-601-8
    1. Human physiology -- Examinations, questions, etc.  I. Title.
II. Series.
    [DNLM: 1. Physiology -- examination questions.   QT 18.2 P282r  1998]
QP40 P37   1998
612'.0076 -- dc21
DNLM/DLC
for Library of Congress                                                     97-48548
                                                                                CIP

**British Library Cataloguing in Publication Data**

Pasley, James N.
    Review questions for human physiology. – (Review questions series)
    1. Human physiology – Examinations, questions, etc.
    I. Title
    612'.0076

    ISBN 1-85070-601-8


This edition published 1998

Printed in the USA

# CONTENTS

Numbers in parentheses indicate number of questions available.

# ACKNOWLEDGEMENTS

The author wishes to thank Dr Robert Skinner, Department of Anatomy, UAMS, for his assistance with the Neurophysiology section and other faculty members of the Department of Physiology and Biophysics, UAMS, for their comments and criticisms. In addition, the author thanks Ms Stacy Major and Ms Tamara Williams for secretarial support and the Staff at Parthenon Publishing, particularly Mr Nat Russo and Ms Roseann Caserio, for their assistance and encouragement.

# INTRODUCTION

Question review is an important component of any discipline of study. The study of human physiology is no exception. These *Review Questions in Human Physiology* have been constructed according to the format used in the United States Medical Licensing Examinations (USMLE), Steps 1, 2, & 3. The question items used include positively worded and negatively worded one-answer multiple-choice questions containing five foils each and extended matching questions containing a minimum of seven and a maximum of 26 foils at the end of each section. Use of this book will provide a learning tool to strengthen the student's ability to apply and integrate the physiological principles necessary for mastery of this important basic science.

# SECTION 1:   GENERAL/CELL PHYSIOLOGY

1.001  Erythrocytes have a normal life span of about 120 days. Normal platelet survival is about:
  A. 1-2 days
  B. 3-4 days
  C. 5-10 days
  D. 10-20 days
  E. unknown

C. is correct
Normal platelet survival is between 9-10 days.

1.002  For each molecule of glucose utilized, the number of ATPs generated during anaerobic glycolysis is:
  A. 1
  B. 2
  C. 3
  D. 4
  E. 5

B. is correct
The net moles of ATP gained from each molecule of glucose subjected to glycolysis in the absence of oxygen are
+ 4 ATP - 2 ATP = 2 ATP.

1.003  Within mammalian cells, adenosine triphosphate (ATP) is distributed:
  A. only in the cell membrane
  B. only in the mitochondria
  C. only in the nucleus
  D. only in the nucleolus
  E. in all cellular structures

E. is correct
ATP is necessary for many cellular functions such as maintaining ion gradients.

1.004  Calcium concentration of relaxed skeletal muscle in the vicinity of thick and thin filaments is low ($\sim10^{-7}$M) because:
  A. calcium is in electrochemical equilibrium across the plasma membrane
  B. troponin binds significant amounts of calcium
  C. circulating red blood cells actively take up calcium which sets up a favorable concentration gradient for depletion of skeletal muscle calcium stores
  D. calcium is actively transported into sarcoplasmic reticulum as well as outside the cell
  E. calcium channels remain closed in quiescent (relaxed) skeletal muscle

D. is correct
Following muscle contraction $Ca^{++}$ is pumped into the SR and outside of the cell. This lowers cytoplasmic concentrations of $Ca^{++}$ in the cell, thus, preventing contraction.

1.005  The resting membrane potential of a typical excitable cell:
  A. is measured as 0 mV immediately upon inhibiting $Na^+$ $K^+$-ATPase activity
  B. is greatly altered if extracellular $Na^+$ is increased to 130 mM
  C. usually equals the equilibrium potential of $Na^+$
  D. is dependent upon the permeability of $K^+$
  E. is dependent upon the permeability of $K^+$ being greater than that of $Na^+$

E. is correct
Potassium and sodium ions can leak through the plasma membrane via "leaking channels": Potassium is 100 x more permeable than sodium.

1.006  The action potential of skeletal muscle:
    A. results in a transient reversal of the concentration gradient for $Na^+$
    B. declines in amplitude as it moves along the surface of the muscle fiber
    C. ends with an outward movement of $K^+$
    D. begins with an outward movement of $Na^+$
    E. is associated with increased acetylcholine receptors within the neuromuscular junction

**C. is correct**
The action potential ends with the rapid diffusion of potassium to the exterior - repolarization.

1.007  Which of the following observations is most consistent with the fluid mosaic model of the plasma membrane?
    A. acetylcholine receptors in skeletal muscle are localized almost exclusively to the neuromuscular junction
    B. by electron microscopy, secretory granules can be visualized in the terminal region of most axons
    C. the cholesterol content of the plasma membrane varies depending on cell type
    D. the plasma membrane is relatively impermeable to ions
    E. none of the above

**E. is correct**
The fluid mosaic model assumes that proteins are embedded at irregular intervals in the membrane and that they freely diffuse through the lipid bilayer.

1.008  Which of the following observations is not consistent with the movement of substance Y across a membrane by active transport?
    A. increasing the temperature resulted in an increase in the rate of transport of substance Y
    B. a second substance with a similar molecular structure to substance Y competitively inhibited transport of substance Y
    C. as the extracellular concentration of substance Y increased, the rate of transport of substance Y increased in a linear fashion
    D. addition of a metabolic poison reduced transport of substance Y across the membrane
    E. after careful measurement, the concentration of substance Y was found to be higher outside the cell than inside the cell

**C. is correct**
Active transport is not dependent upon concentration gradients. Active transport moves substances from an area of high to low concentration or from a low to high concentration.

1.009  Which of the following statements provides the reason why a given skeletal muscle can maintain a submaximal contractile state for long periods of time without a decrease in tension due to fatigue?
    A. individual motor units fire asynchronously
    B. repetitive stimulation increases TCA cycle and oxidative phosphorylation activity
    C. passive tension is provided by elastic elements in muscle (e. g., tendons and connective tissue)
    D. phosphofructokinase activity is increased in actively contracting muscle
    E. none of the above statements provide the explanation for the observation.

**A. is correct**
Muscle fibers contract in groups called muscle units. The muscle fibers of a muscle unit are innervated by a motor neuron - this is known as a motor unit. When individual motor units fire asynchronously - some motor units contract while others remain at rest.

1.010 In skeletal muscle, tropomyosin functions primarily to:
   A. release calcium upon initiation of contraction
   B. form the crossbridge between actin and myosin
   C. provide structural rigidity to the thin filament
   D. cover the sites on g-actin that bind myosin
   E. hydrolyze ATP to ADP and $P_i$

**D. is correct**
Tropomyosin covers the sites on g-actin that bind myosin.

1.011 Which of the following is correct for rough endoplasmic reticulum?
   A. concentrates calcium in smooth muscle
   B. is involved in steroid biosynthesis in the adrenal cortex
   C. is involved in phospholipid biosynthesis in nearly all cells
   D. is involved in protein synthesis in nearly all cells
   E. is the major source of sodium in the cell

**D. is correct**
The endoplasmic reticulum is a complex of tubules in the cell cytoplasm. Proteins synthesized in the endoplasmic reticulum are sorted by the Golgi apparatus and accumulated in vesicles or storage granules.

1.012 Steroid hormones are believed to enter target cells via:
   A. facilitated diffusion (carrier-mediated transport)
   B. receptor-mediated endocytosis
   C. cholesterol-lined pores in the plasma membrane
   D. simple diffusion
   E. active transport

**D. is correct**
Steroid hormones are lipophilic and enter the cell through the plasma membrane by simple diffusion.

1.013 Active transport can occur:
   A. only in excitable cells and certain epithelial-like cells
   B. when a solute moves from a region of high electrochemical potential to a region of low electrochemical potential
   C. in the absence of an energy source
   D. if the solute transport across the cell membrane is not driven by the hydrolysis of ATP
   E. without oxygen

**B. is correct**
Active transport can transport a solute from an area of low to high electrochemical potential or from an area of high to low electrochemical potential.

1.014 In exercising skeletal muscle, ATP is used for which of the following?
   A. to provide energy for the thick and thin filaments to slide past each other
   B. to increase the affinity between actin and myosin, and thereby allow cyclic interaction of the two proteins
   C. calcium secretion out of the cell
   D. conduction of the action potential throughout the t-tubule system
   E. to release $Ca^{++}$ from the sarcoplasmic reticulum

**A. is correct**
ATP provides the energy source so that thick and thin muscle filaments can slide past each other.

1.015  All of the following are correct for action of curare on neuromuscular transmission, EXCEPT:
- A. irreversible binding to the acetylcholine receptor
- B. inhibition of calcium uptake by sarcoplasmic reticulum
- C. opening of sodium channels allowing sodium to flow down its concentration gradient
- D. inhibition of acetylcholinesterase activity
- E. increasing calcium uptake by sarcoplasmic reticulum

E. is correct
Curare blocks muscle contraction by binding to and preventing normal function of the acetylcholine receptor.

1.016  Which of the following statements would not be considered consistent with the fluid mosaic model of membrane structure?
- A. cellular membranes are fluid structures in which proteins are free to diffuse in the plane of the membrane
- B. in an aqueous environment the most energetically stable configuration for phospholipids is to form structures that allow the fatty acid chains to be kept away from contact with $H_2O$
- C. integral or intrinsic membrane proteins are embedded in the phospholipid bilayer
- D. cells consume relatively large amounts of ATP in order to maintain the phospholipid bilayer
- E. integral proteins have hydrophobic domains

D. is correct
The phospholipid bilayer is the most energetically favorable conformation. No ATP required.

1.017  Examples of active transport processes include all of the following EXCEPT:
- A. amino acid transport into most cells
- B. transport of glucose across epithelial cells
- C. $Ca^{++}$ transport into sarcoplasmic reticulum
- D. extrusion of $Na^+$ following an action potential
- E. transport of fructose across intestinal epithelial cells

E. is correct
Fructose is transported across intestinal epithelial cells by facilitated diffusion.

1.018  All of the following are involved in determining the electrochemical potential difference of an ion across the membrane EXCEPT:
- A. the concentration gradient of the ion
- B. the valency or charge of the ion
- C. the ambient temperature
- D. the molecular weight of the ion
- E. the magnitude of the voltage difference across the membrane

D. is correct
Clearly, molecular weight is not involved in determining equilibrium potential.

1.019  A motor unit:
- A. is a large group of muscle fibers
- B. contracts as a unit
- C. is a single motor neuron and all the muscle fibers that it innervates
- D. varies in size from muscle to muscle
- E. all of the above

E. is correct
Skeletal muscle fiber contraction groups called muscle units fire asynchonously allowing some to contract and others to rest.

1.020  In skeletal muscle during a muscle contraction:
  A. calcium is not released from the sarcoplasmic reticulum
  B. the hydrolysis of ATP provides the energy required in order for myosin to flex within the "hinge regions"
  C. the rigor complex is broken by ADP binding to myosin
  D. the myosin-ATP complex demonstrates high affinity for actin
  E. f-actin is hydrolyzed to g-actin

B. is correct
The hydrolysis of ATP is required for the production of energy for muscle contraction.

1.021  Fast glycolytic (white) skeletal muscle fiber types:
  A. are abundant in muscles involved in maintenance of posture
  B. have relatively high levels of glycolytic enzymes
  C. have relatively high myoglobin levels
  D. fatigue relatively slowly
  E. have a relatively high density of capillaries

B. is correct
Slow glycolytic (red) skeletal muscle fibers are involved in slow prolonged muscle contraction (e. g. - maintenance of posture). Myoglobin levels are high, therefore, and they have a high density of capillaries that supply abundant oxygen.

1.022  In most excitable cells, $K^+$ rather than $Na^+$ plays the dominant role in determining the resting membrane potential because:
  A. hydrated $Na^+$ is larger than hydrated $K^+$
  B. there are more $K^+$ channels than $Na^+$ channels in the plasma membrane
  C. the plasma membrane is more permeable to $K^+$ than it is to $Na^+$
  D. the intracellular $K^+$ concentration is higher than the intracellular $Na^+$ concentration
  E. the $K^+$ electrochemical potential difference across the membrane is greater than that for $Na^+$

C. is correct
Potassium is a 100x more permeable than sodium.

1.023  Which of the following compounds does not stimulate activity of the rate limiting enzyme in glycolysis, phosphofructokinase?
  A. epinephrine
  B. AMP
  C. ADP
  D. ATP
  E. fructose 6-phosphate

D. is correct
After fructose 6-phosphate is converted to fructose 1,6-biphosphate it is committed to glycoysis. Phosphofructokinase catalyzes this reaction. When ATP levels are high this signals to the cell that ATP is being made faster than it can be consumed and therefore inhibits phosphofructokinase.

1.024  Which statement concerning smooth muscle is true?
  A. all smooth muscle has a basal resting tension or tone upon which contractions are superimposed
  B. smooth muscle responds only to acetylcholine
  C. visceral smooth muscle has well-developed neuromuscular junctions
  D. the latent period in visceral smooth muscle is shorter than that in skeletal muscle
  E. multi-unit smooth muscle develops tension in response to stretch

A. is correct
Smooth muscle exhibits a resting basal tension upon which contractions are superimposed.

1.025  Which statement is true concerning miniature end plate potentials (MEPPS)?
    A. MEPPS often elicit an action potential
    B. MEPPS are due to a defect in the acetylcholine receptor
    C. MEPPS hyperpolarize the skeletal muscle membrane
    D. MEPPS help maintain the integrity of the t-tubule system
    E. MEPPS are due to the spontaneous release of acetylcholine-containing vesicles from the motor nerve terminal

E. is correct
The spontaneous release of small amounts of acetylcholine from the nerve cell membrane causes a minute depolarizing spike called a miniature end-plate potential (MEPP).

**Use the following list to answer questions 26-29.**

    A. potassium
    B. chloride
    C. sulfate
    D. sodium
    E. protein
    F. calcium
    G. bicarbonate

1.026  Principal diffusion anion in interstitial fluid

B. is correct

1.027  Difference in concentration across the capillary endothelium causes concentrations of diffusible ions to differ in plasma and interstitial fluid

E. is correct

1.028  Body content of exchangeable ion is proportional to the volume of extracellular fluid

D. is correct

1.029  Electrolyte with the highest intracellular concentration.

A. is correct

1.030  The absolute refractory period following an action potential is due to:
    A. insufficient neurotransmitter
    B. depletion of intracellular $Na^+$
    C. inactivation of $Na^+K^+$-ATPase
    D. voltage inactivation of the $Na^+$ channel
    E. none of the above

D. is correct
$Na^+$ channels became inactivated when the membrane is depolarized. The $Na^+$ channels will not open until the potential returns to the resting membrane potential.

1.031  The first event during an action potential is:
    A. an increase in $K^+$ conductance
    B. a decrease in $K^+$ conductance
    C. an increase in $Na^+$ conductance
    D. a decrease in $Na^+$ conductance
    E. depletion of intracellular $Na^+$

C. is correct
Events that occur during an action potential include: 1) an increase $Na^+$ conductance; 2) a decrease $Na^+$ conductance; 3) increased $K^+$ conductance; and 4) decreased $K^+$ conductance.

**Use the following list to answer questions 32-34:**

    A. intracellular fluid
    B. extracellular fluid
    C. plasma
    D. transcellular fluid
    E. total body water

1.032 Estimated by subtracting the volume of distribution of inulin from the volume of distribution of tritiated water

A. is correct
ICF = TBW - ECG ; ITSF = ECF - PV

1.033 Volume in a normal young adult male is generally > 50% of the body weight.

E. is correct
In the average young adult male, 60% of the body weight is water, 15% is fat, 18% is protein and 7% mineral.

1.034 Measured by determining the volume of distribution of $I^{131}$ or $T^{1824}$.

C. is correct
$PV = I^{131}$ or $T^{1824}$

Use the following figure to answer questions 35-37.

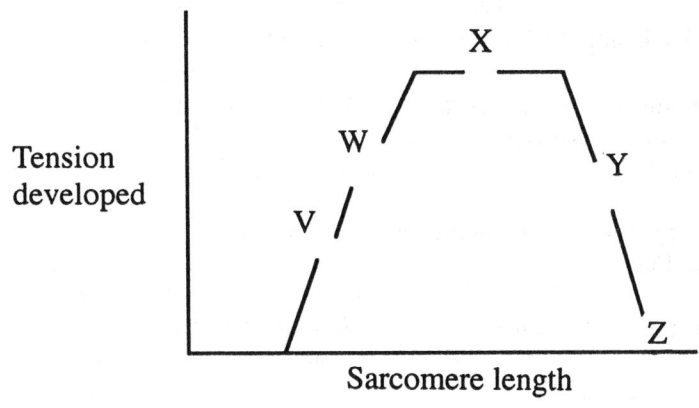

1.035 At which point on the length tension curve is it possible to form the maximum number of crossbridges?
    A. V
    B. W
    C. X
    D. Y
    E. Z

C. is correct
Point X represents the optional length of the sarcomere that results in maximal tension being produced upon stimulation.

1.036 At which point on the length tension curve is there no overlap of thick and thin filaments?
    A. V
    B. W
    C. X
    D. Y
    E. Z

E. is correct
Point Z is where crossbridge formation is not possible since the thick and thin filaments do not overlap.

1.037  Which of the following explains the decreased tension development between points W and V?
   A. folding of the thick filament
   B. overlap of adjacent thin filaments
   C. dissociation of f-actin into g-actin
   D. increased inhibition of crossbridge formation by the troponin-tropomyosin complex
   E. disruption of t-tubule function

B. is correct
Between these points the sarcomere has shortened to a point where thin filaments overlap.

1.038  Which of the following is correct for gamma motoneurons of the spinal cord?
   A. innervate skeletal muscle fibers
   B. innervate antagonistic muscles only
   C. innervate intrafusal muscle fibers
   D. stimulate Ib afferent fibers
   E. inhibit skeletal muscle contraction

C. is correct
The gamma motoneurons are located in the ventral horn of the spinal cord and innervate the intrafusal fibers of the muscle spindles.

**Use the following table to answer the question 39.**

|     | Plasma osmolality | ECFV | ICFV |
| --- | --- | --- | --- |
| A. | increase | increase | decrease |
| B. | decrease | increase | increase |
| C. | decrease | decrease | increase |
| D. | increase | decrease | decrease |
| E. | no change | increase | no change |

ECFV = volume of extracellular fluid; ICFV = volume of intracellular fluid

1.039. Intravenous infusion of 1 liter of isotonic sodium chloride solution will result in which of the above:

E. is correct
This is an example of isotonic overhydration where volume of fluid outside of the cell gets larger (ECFV) with no change in volume of intracellular fluid (ICFV); since the fluid is isotonic there is no change in plasma osmolality.

**Use the following graph to answer question 40**

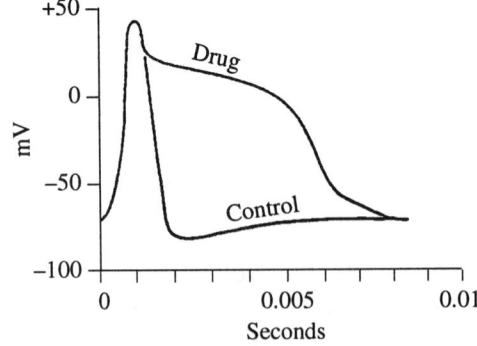

1.040 The graph above illustrates the effects of tetraethylammonium (TEA) on the action potential of an axon. The most likely action of TEA is to:
    A. open K⁺ gates, causing membrane permeability to K⁺ to increase during the action potential
    B. block K⁺ preventing membrane permeability to K⁺ from increasing during the action potential
    C. open Na⁺ gates, causing membrane permeability to Na⁺ to increase during the action potential
    D. close Na⁺ gates, preventing membrane permeability to Na⁺ from increasing during the action potential
    E. block the Na⁺ - K⁺ active pump

B. is correct
On the control curve the upward depolarization is caused by the influx of Na⁺ into the axon. The downward repolarization is due to the reflux of K⁺ out of the axon. On the "dry" curve there is no downward repolarization. Therefore, TEA decreases membrane permeability to K⁺.

**Use the following diagram to answer question 41**

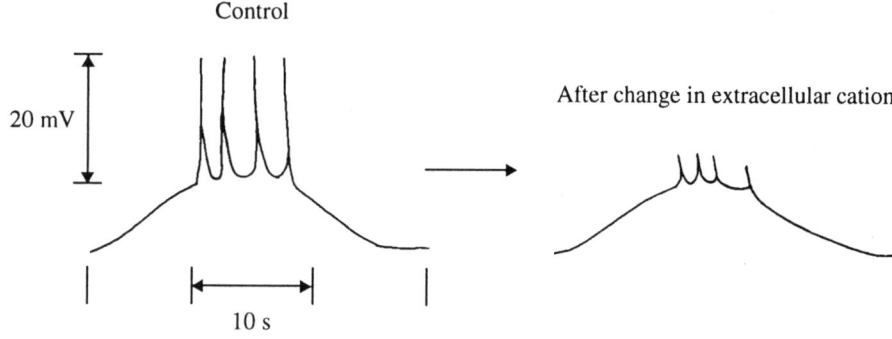

1.041 The above diagram shows the effect of a change in cationic composition of a bathing fluid on the intracellular potential of uterine smooth muscle. Of the following, what change occurs to produce the observed effects?
    A. an increase in Na⁺
    B. a decrease in K⁺
    C. an increase in Ca⁺⁺
    D. a decrease in Ca⁺⁺
    E. an increase in K⁺

A. is correct
An increase in sodium in the extracellular solution will cause the intracellular potential of the uterine smooth muscle to be more negative.

**Use the following diagram to answer question 42.**

Hypertonic solutions of either KCl or sucrose are added to an erythrocyte suspension. Changes in cell volume as a function of time after addition are recorded as follows:

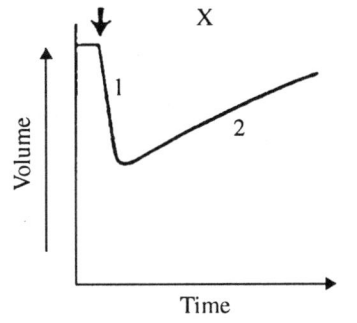

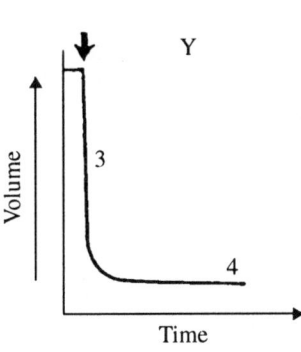

1.042  From the above information, which of the following statements is correct?
  A.  curve Y represents a KCl experiment
  B.  portion 2 of curve X represents the efflux of water
  C.  portion 3 of curve Y represents the influx of water
  D.  erythrocytes are impermeable to KCl
  E.  erythrocytes are more permeable to water than to either solute

E. is correct
Curve Y represents the sucrose solution, portion 2 of curve X represents the influx of water; portion 3 of curve Y represents the efflux of water; erthyrocytes are very permeable to KCl - see curve Y. Thus, by elimination, erythrocytes are more permeable to water than either sucrose or KCl.

**Use the following diagram to answer question 43.**

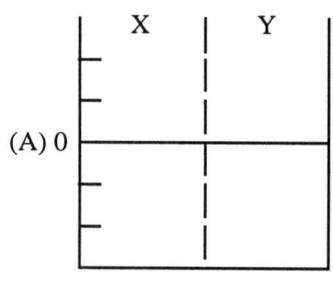

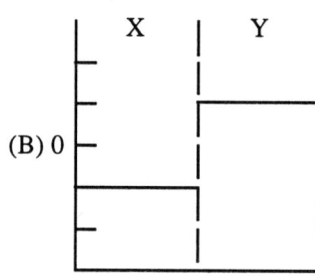

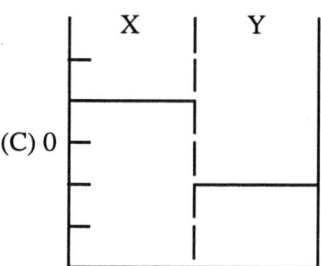

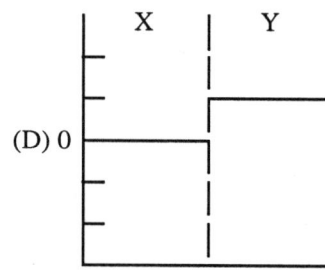

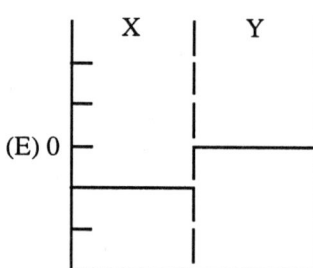

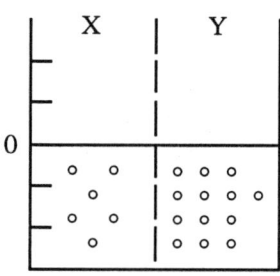

1.043  Solutions X and Y are separated by a semi-permeable membrane. The concentration of solutes at time 0 is depicted above. At equilibrium, the volumes of compartments X and Y will be:

B. is correct
Due to the osmotic gradient, water will move from compartment X to compartment Y until equilibrium is reached. Only choice B depicts the same volume of water as at time 0.

**Use the following graphs to answer questions 44-46.**

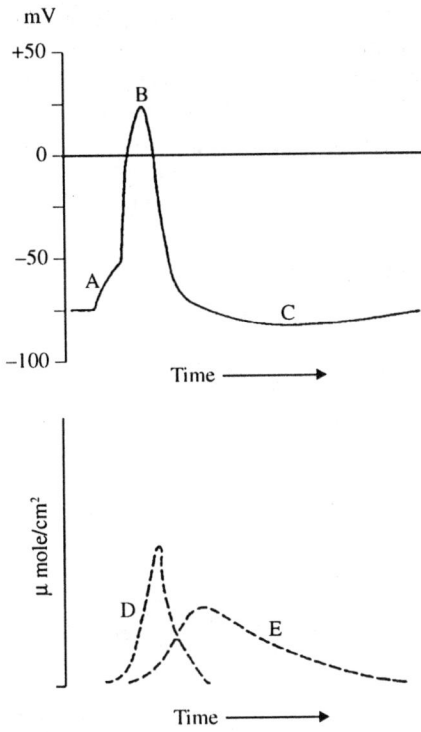

The above diagrams depict the action potential of a spinal motoneuron in a small mammal as recorded with an intracellular microelectrode (solid line) and events directly related to it (broken lines). For each numbered statement below, select the one letter designating the part of one of the diagrams that matches it most closely.

1.044  Na⁺ equilibrium potential               B. is correct

1.045  K⁺ conductance                          E. is correct

1.046  K⁺ equilibrium potential                C. is correct

**Use the following diagram to answer questions 47 and 48**

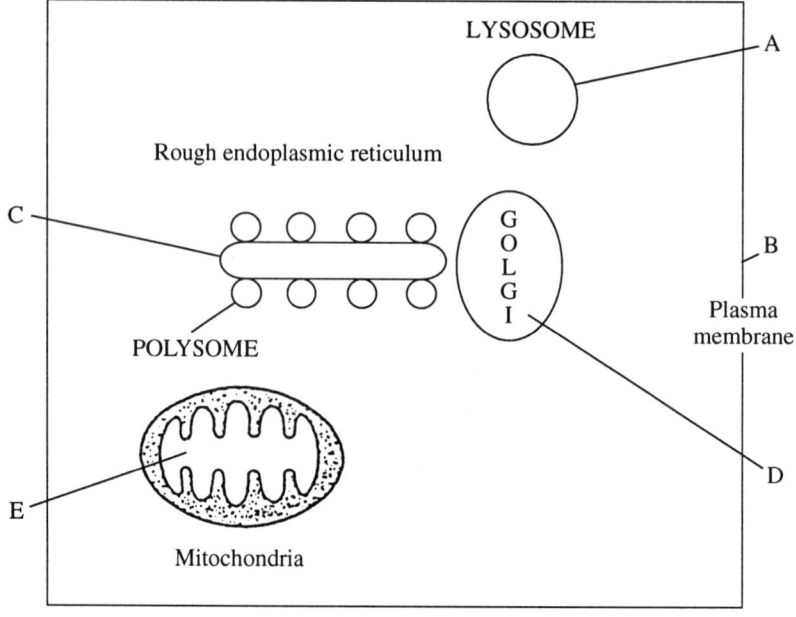

1.047  Contains DNA

E. is correct
Mitochondria have circular DNA and the ability to synthesize RNA from it.

1.048  Site of assembly of secretory vacuoles

D. is correct
Peptide hormones are packaged within Golgi - derived membranes and are secreted at the cell surface by exocytosis.

1.049  A normal subject was administered 300 mg of Evan's Blue (T1824) intravenously. After equilibration the plasma concentration of the dye was 10 mg/100 ml. If the hematocrit (Hct) was 40%, blood volume would be closest to:
   A.  3. 1 liters
   B.  4. 0 liters
   C.  5. 0 liters
   D.  6. 3 liters
   E.  7. 5 liters

C. is correct
BV = PV 100/100 - Hct; BV = 3l 100/100-40; BV = 5. 0 l

1.050  The tissue that contains the largest amount of the body's non-exchangeable sodium is:
   A.  blood
   B.  skeletal muscle
   C.  gastrointestinal tract
   D.  bone
   E.  skin

D. is correct
Bone is the tissue that contains the largest amount of non-exchangeable sodium.

1.051  All of the following will increase the insensible loss of water, EXCEPT:
   A.  decrease in osmolality of body water
   B.  decrease in alveolar ventilation (respiration)
   C.  decrease in humidity of inspired air
   D.  increase in body temperature
   E.  none of the above

B. is correct
Increased respiration rates increase the insensible loss of water via enhanced expiration.

**Use the following diagram to answer questions 52 and 53.**

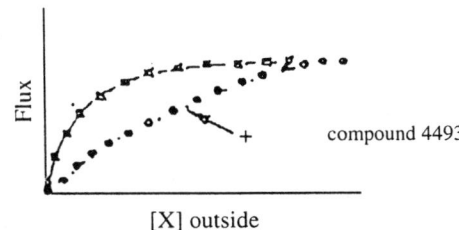

1.052  From the above figure, transport of substance X into adipocytes is most likely by:
   A.  simple diffusion
   B.  facilitated diffusion
   C.  active transport
   D.  co-transport
   E.  insufficient information contained within the figure to answer the question

E. is correct
Simple diffusion does not reach saturation as the graph shows saturation. Both facilitated diffusion and active transport display saturability. Two substances are transported in co-transport. Thus, there is insufficient information to answer this question.

1.053  From the above figure, compound 4493 most likely inhibits the uptake of substance X by:
   A.  inhibiting $Na^+K^+$-ATPase activity
   B.  inhibiting mitochondrial oxidative phosphorylation
   C.  competitive inhibition
   D.  non-competitive inhibition
   E.  insufficient information to answer the question

C. is correct
The figure is representative of classic competitive inhibition. The rate of transport is dependent upon the concentration of both the inhibitor and the substrate. In non-competitive inhibition the rate of transport is dependent upon only the concentration of inhibitor. In competitive inhibition increasing the concentration of substrate will overcome the inhibitor's effect.

1.054  Which of the following characteristics do not affect the osmotic pressure exerted by a solute in solution?
   A.  temperature of the solution
   B.  concentration of the solute in the solution
   C.  degree of dissociation of the solute in the solution
   D.  molecular weight of the solute
   E.  none of the above

D. is correct
According to Van't Hoff's law, molecular weight is not a valuable in determing osmotic pressure.

1.055  In the plasma membrane of the hepatocyte, the arrangement of the phospholipids is best described by which of the following statements?
  A. randomly
  B. a rigid bilayer
  C. a fluid bilayer
  D. the relatively hydrophilic (water-loving) fatty acid chains (tails) pointing inward
  E. extensive areas predominantly consisting of cholesterol with occasional patches composed of glycolipids

C. is correct
Phospholipids from a bilayer have hydrophilic (water loving) heads on the exterior and hydrophobic (water-fearing) fatty acid chains on the interior away from the water.

1.056  Which statement concerning smooth muscle is true?
  A. smooth muscle has a shorter latent period than skeletal muscle
  B. smooth muscle responds only to acetylcholine
  C. visceral smooth muscle has a well-developed neuromuscular junction
  D. multi-unit smooth muscle develops tension in response to stretch
  E. all smooth muscle has a basal resting tension or tone

E. is correct

1.057  Which statement regarding skeletal muscle fiber types is true?
  A. compared to type IIB (white) muscle fibers, type I (red) muscle fibers have a high glycolytic capacity
  B. the motor nerve has no influence on the muscle fiber type that it innervates
  C. compared to type I (red) muscle fibers, type IIB (white) muscle fibers have a relatively increased amount of myoglobin
  D. prolonged exercise increases the number of type I (red) muscle fibers in the quadriceps muscle
  E. type I (red) muscle fibers have a higher oxidative capacity than type IIB (white) muscle fibers

E. is correct
Type I red muscle fibers are slow twitch fibers that have large amounts of mitochondria, their metabolism is primarily oxidative and they are relatively fatigue-resistant.

1.058  ATP consumption in contracting skeletal muscle increases as:
A. blood flow to the muscle decreases
B. the magnitude of the action potential increases
C. the load on the muscle increases
D. the velocity of shortening decreases
E. none of the above

C. is correct
As muscle load increases work increases. The energy to perform work is provided by ATP. Therefore, ATP consumption increases.

1.059  The skeletal muscle action potential:
  A. is not essential for contraction to occur
  B. has a prolonged plateau phase
  C. spreads inward to all parts of the muscle via the t-tubule system
  D. begins with an inward movement of $K^+$ ions
  E. stimulates the immediate uptake of $Ca^{++}$ by the sarcoplasmic reticulum

C. is correct
The sarcolemma dips deep into the interior of the cell in the form of finger-like projections called t-tubules which allows rapid spread of the action potential throughout the cell.

1.060  Skeletal muscle can maintain a submaximal contraction for a long period of time without a decrease in tension due to fatigue because:
    A. passive tension increases as tendons and connective tissues are stretched
    B. lactic acid is metabolized to $CO_2$ and $H_2O$ via the Kreb's cycle
    C. individual motor units are recruited and contract asynchronously
    D. skeletal muscle can be tetanized
    E. crossbridge formation depends upon an adequate supply of ATP

C. is correct
One-way increased tension development in skeletal muscles is attained by recruitment of motor units allowing a continuous and smooth movement because motor units fire asynchronously.

1.061  If a drug that blocks sodium channels is applied to ventricular muscle, the phase of the action potential that will be most affected is:
    A. 0 (rapid upstroke)
    B. 2 (plateau)
    C. 3 (rapid repolarization)
    D. 4 (resting or diastolic)
    E. none of the above

A. is correct
0 phase or rapid upstroke of the action potential is where sodium influx occurs.

**Use the following graph to answer questions 62 and 63.**

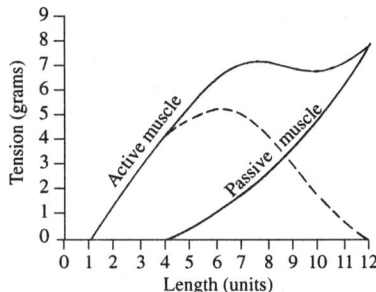

The graph above illustrates the isometric length-tension relations of a muscle fiber before stimulation (passive) and after maximal stimulation (active). The dashed curve is the difference between these two relations.

1.062  If the muscle is preloaded with a 4 gm weight and then tetanically stimulated to contract isotonically, it will shorten by approximately:
    A. 1 unit
    B. 3 units
    C. 5 units
    D. 7 units
    E. 9 units

E. is correct

1.063  The maximum tension that can be generated by the contractile apparatus is observed at a fiber length of:
   A.  2 units
   B.  4 units
   C.  6 units
   D.  8 units
   E.  12 units

C. is correct

**Use the following information to answer question 64.**

osmolarity = $\Phi ic$
for $CaCl_2$: $\Phi = 0.94$
   i = # of ions formed upon dissociation
   c = molar concentration of solute

1.064  The osmolarity of a 0.130 M $CaCl_2$ solution (1 liter total volume) is closest to:
   A.  122 milliosmoles/liter
   B.  144 milliosmoles/liter
   C.  367 milliosmoles/liter
   D.  390 milliosmoles/liter
   E.  410 milliosmoles/liter

C. is correct
The osmolarity of a solution is the number of osmoles per liter of solution. One osmole is equal to 1000 milliosmoles (mOsm).
Osmolarity = $\Phi ic$. $\Phi = 0.94$; i = 3; c = 0.130
$\Phi ic = (0.94)(3)(0.130 M)$
$\Phi ic = 0.367$ osmoles/liter
$\Phi ic = 367$ milliosmoles/liter

1.065  The rapid phosphorylation by hexokinase of glucose upon entry into a muscle cell is physiologically significant for which of the following reasons?
   A.  insulin and exercise both enhance glucose uptake probably by increasing hexokinase activity
   B.  phosphorylated glucose efficiently enters the mitochondria where it is completely metabolized via the Kreb's cycle
   C.  intracellular glucose concentration is maintained at a relatively low level thereby maintaining a favorable concentration gradient for glucose entry into the cell
   D.  the hexokinase-catalyzed phosphorylation of glucose in muscle cells is readily reversible so that the muscle is able to help maintain blood glucose levels during times of starvation
   E.  high levels of glucose-6-phosphate promote breakdown of glycogen

C. is correct
Glucose is taken up by most cells by facilitated diffusion, therefore, the flux of glucose is related to the concentration gradient. Glucose is phosphorylated decreasing the glucose concentration in the cell making glucose-6-phosphate. This creates a favorable environment for glucose influx.

1.066  Which of the following statements concerning the skeletal muscle nicotinic acetylcholine receptor is false?
   A.  the receptor besides binding acetylcholine also serves as a $Na^+$ channel
   B.  in normally innervated skeletal muscle, the receptor is found distributed evenly over the entire surface of individual muscle cells
   C.  patients with myasthenia gravis frequently have circulating antibodies against this receptor
   D.  a cobra snake bite is frequently fatal because the venom contains $\alpha$-bungarotoxin which binds irreversibly to the nicotinic acetylcholine receptor
   E.  the receptor is a large glycoprotein composed of several subunits

B. is correct
The neuromuscular junction (or ACh receptor) is located near the middle of the muscle fibers. There is only one neuromuscular junction per fiber.

1.067   In skeletal muscle during a contraction cycle:
   A.   extracellular calcium crosses the sarcolemma and binds to tropomyosin
   B.   myosin-ATP complex has a high affinity for actin
   C.   the rigor complex is broken by the binding of ADP to myosin
   D.   f-actin depolymerized to g-actin
   E.   the hydrolysis of ATP provides the energy required in order to myosin to flex in its hinge regions

E. is correct
ATP hydrolysis provides the energy for myosin to flex in its hinge regions.

**Use the following information to answer questions 68 and 69.**

An unusual cell is found to have a resting membrane potential of -50 mV (inside negative with respect to outside) and its intracellular $Mg^{++}$ concentration is 25 mM. The extra-cellular $Mg^{++}$ concentration is 5 mM.

m.p. = 50 mV

$[Mg^{++}]i = 25$ mM

$[Mg^{++}]o = 5$ mM

$$voltage \ = \ \frac{-60}{z} \ log \frac{[x]i}{[x]o} \ mV$$

log 25 = 1.39   log 5 = 0.699
log 20 = 1.30   log 0.2 = –0.699
z = valence

1.068   What is the equilibrium potential for $Mg^{++}$ across this cell's plasma membrane?
   A.   0 mV
   B.   50 mV
   C.   -50 mV
   D.   -21 mV
   E.   21 mV

D. is correct

$$E_{Mg} \ = \ \frac{-60}{z} \ log \ \frac{25}{5} \ mV$$

$$\frac{-60}{2} \ log \ 5 \ mV \ = \ -20.97 \ or \ -21 \ mV$$

Inside negative with respect to outside.

1.069   If a metabolic poison was administered such that all active transport processes are inhibited 100%, in which direction would the net movement of $Mg^{++}$ be?
   A.   inward
   B.   outward
   C.   no net movement because $Mg^{++}$ is in equilibrium
   D.   initially inward, but then outward
   E.   insufficient information has been given to answer the question

A. is correct
Because $Mg^{++}$ is a positive ion and the inside potential is negative with respect to outside, $Mg^{++}$ would move inward.

**Use the following graph to answer question 70.**

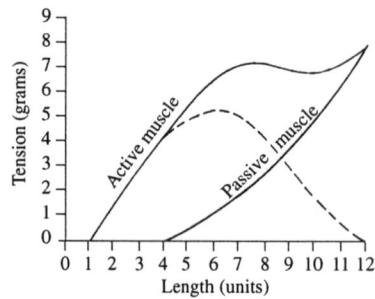

1.070  The segment on the length-tension relationship over which skeletal muscle functions with greatest active tension development:
  A.  1-2
  B.  2-3
  C.  3-4
  D.  4-5
  E.  2-5

C. is correct
The top curve represents total tension; the middle curve active tension; and the bottom curve passive tension. The sarcomere length at which the greatest active tension is developed is approximately (segment 3-4.)

1.071  The percentage of ATP produced aerobically by exercising muscle during 15 minutes of submaximal exercise would be expected to:
  A.  remain constant, approximately 20% of total ATP production
  B.  increase
  C.  decrease
  D.  initially increase, but then decrease to approximately 50% of total ATP production
  E.  exceed the $O_2$ debt by 10%

C. is correct
Oxidative phosphorylation is a relatively slow process which cannot meet ATP demands in rapidly contracting muscle.

1.072  Which of the following drugs might be effective in reducing the magnitude of strong muscle spasms that are asserted with lower back pain in a 55-year-old man?
  A.  a drug that inhibited $Ca^{++}$ release from sarcoplasmic reticulum
  B.  a drug that inhibited myosin light chain kinase
  C.  a drug that blocked $Ca^{++}$ channels in the cell membrane
  D.  an acetylcholinesterase inhibitor
  E.  an inhibitor of skeletal muscle glucose uptake

A. is correct
By inhibiting $Ca^{++}$ release from the SR, $Ca^{++}$ will never bind with troponin C and the active sites on actin will not be exposed.

1.073  Increased total tension developed by skeletal muscle can be induced by:
  A.  recruitment of additional motor units
  B.  increased frequency of motor nerve stimulation
  C.  stretching the muscle to twice its resting length prior to contraction
  D.  A and B only
  E.  A, B and C

D. is correct
Increasing the number of motor units with a greater number of muscle fibers as well as increasing the frequency of motor nerve stimulation can lead to greater tension, summation and tetanus.

1.074  Ca$^{++}$ induced activation of skeletal muscle contraction involves which molecular mechanism:
A. binding of Ca$^{++}$ to calmodulin
B. binding of Ca$^{++}$ to g-actin
C. enhanced activity of phosphofructokinase
D. activation of myosin ATPase activity
E. binding of Ca$^{++}$ to troponin

E. is correct
Calcium binds to troponin C and causes tropomyosin to slide into grooves of f-actin and thereby exposing the myosin binding sites on g-actin molecules.

1.075  A smooth, sustained, submaximal contraction of skeletal muscle is dependent upon:
A. the presence of both type I and type IIB fiber types
B. the ability to recruit individual motor units
C. identical twitch durations of all fiber types
D. rapid achievement of tetanus
E. synchronization of each crossbridge cycle

B. is correct
Increased total tension is provided by recruitment of more motor units.

1.076  The almost immediate sensation of pain that follows painful trauma seems to be carried to the central nervous system by a different set of fibers from those that carry more delayed unpleasant painful sensation. The first sharp sensations of pain are carried by:
A. type A (delta) fibers
B. type B fibers
C. type C fibers
D. type A (alpha) fibers
E. type D fibers

A. is correct
Type A (delta) fibers are responsible for transmitting the immediate sensation of pain.

1.077  The volume of distribution is greatest for which of the following substances?
A. albumin
B. sodium
C. inulin
D. chloride
E. potassium

E. is correct
Potassium exhibits the greatest volume of distribution.

1.078  The average daily water requirement under basal conditions for a 70 kg (154 lb) male is:
A. 800 - 1,500 ml
B. 1,800 - 3,000 ml
C. 3,000 - 4,000 ml
D. 5,000 - 6,000 ml
E. 6,000 - 8,000 ml

B. is correct
The average daily intake of water for a 70 Kg male equals approximately 2. 6 l; 1. 5 l ingested, 0. 7 l in food and 0. 4 l from metabolic process.

1.079  The blood buffers consist of plasma proteins, hemoglobin, bicarbonates and inorganic phosphate. All of the following would increase the effectiveness of the imidazole group of histamine in hemoglobin to buffer hydrogen ions, EXCEPT:
A. increased PCO$_2$
B. increased pH
C. decreased pH
D. increased (H$_2$CO$_3$)
E. deoxygenation of oxyhemoglobin

B. is correct
The pK$_a$ of the imidazole group of histamine is 6. 0 while physiological pH is approximately 7. 4. Thus, in order for imidazole to be an effective buffer, the pH must be lowered not increased.

1.080  Cell membranes contain all of the following, EXCEPT:
- A. receptors for neurotransmitter
- B. receptors for glucagon
- C. adenylate cyclase
- D. the sodium-potassium-ATPase pump
- E. receptors for steroid hormones

E. is correct
Receptors for steroid hormones are located in the cytoplasm or nucleus not in cell membranes. Steroid hormones are lipophilic and can, therefore, pass through cell membranes.

1.081  Which of the following would produce effects similar to those of myasthenia when administered to a normal individual:
- A. black widow spider toxin
- B. neostignine
- C. curare
- D. 'nerve gas'
- E. a Barry Manilow recording

C. is correct
Curare and other drugs related to tubocumaine are antagonists of acetylcholine at the motor end plate. Because they bind to the acetylcholine receptor, the muscle cannot be stimulated to contract.

1.082  The ratio of the osmotic pressure of plasma to the osmotic pressure of interstitial fluid equals:
- A. 1. 0 at all times
- B. slightly greater than 1. 0
- C. slightly less than 1. 0
- D. a wide fluctuation exists
- E. none of the above

B. is correct
Recall that capillary osmotic pressure is much larger than tissue osmotic pressure. Therefore,

$$\frac{COP}{TOP} = > 1.0$$

**Use the following for questions 83 and 84.**

- A. plasma
- B. extracellular fluid
- C. interstitial fluid
- D. intracellular fluid
- E. transcellular fluid

1.083  Volume is equal to the difference in volumes of distribution of radioactively-labeled sulfate and T-1824:

C. is correct
Interstitial fluid volume can be measured by subtracting the volume of distribution of radioactively-labeled sulfate from the volume of distribution of T-1824 (Evans Blue).

1.084  Fluid formed by transport activity of epithelial tissue:

E. is correct
The total volume of TSF is about 1 liter compared to TBW (40-42 liters). The composition of components of TSF can vary from normal ECF. For example, changes in composition of ESF can occur when excessive losses of gastric juice containing $H^+$ and $Cl^-$ can occur after prolonged vomiting.

**Use the following to answer questions 85 and 86.**

    A. potassium
    B. chloride
    C. sodium
    D. protein
    E. sulfate

1.085  Concentration in plasma tends to increase when pH of extracellular fluid decreases (hydrogen ion concentration increases).

A. is correct
Since most potassium is in ICF, a shift of a small fraction of potassium from ICF to ECF can cause large changes in plasma levels.

1.086  A deficit may cause a reduction in blood volume:

C. is correct
Body sodium content is correlated with the volume of ECF. Thus, excesses and deficits are associated with edema and hypovolemia, respectively.

**Use the following to answer questions 87 and 88.**

    A. sensible perspiration
    B. pancreatic juice
    C. gastric juice
    D. insensible perspiration
    E. fecal water

1.087  Water loss without loss of solutes.

D. is correct
Insensible water loss is water lost through the lungs and skin.

1.088  Bicarbonate concentration is much higher than in extracellular fluid.

B. is correct
Pancreatic juice has a high bicarbonate content and a low chloride content.

**Use the following to answer questions 89 and 90.**

    A. addition of 300 millimoles of NaCl
    B. loss of 300 millimoles of NaCl
    C. addition of 1 liter of water
    D. loss of 1. 5 liters of water

(Changes refer to a normal 70 kg subject)

1.089  Increase in plasma osmolality, increase in volume of extracellular fluid, and decrease in volume of intracellular fluid.

A. is correct
The addition of 300 millimoles of NaCl will result in a decrease in intracellular fluid along with an increase in osmolality and volume of extracellular fluid.

1.090  Increase in plasma osmolality and decrease in the volume of extra- and intracellular fluids.

D. is correct
Water loss will decrease extracellular and intracellular fluid volumes along with an increase in plasma osmolality.

**Use the following list to answer questions 91-96.**

    A.  f-actin
    B.  g-actin
    C.  α-actinin
    D.  myosin
    E.  ADP
    F.  troponin T
    G.  troponin C
    H.  troponin I
    I.  tropomyosin
    J.  ATP
    K.  calcium

Match the muscle protein in the list of answers below with the most appropriate statement concerning this protein's physiologic role in muscle contraction in the items below. Answers may be used once, more than once, or not at all.

1.091  Regulatory muscle protein which lies in the groove of the F-actin helices.

    I. is correct

1.092  Binds actin filaments to Z-disc.

    C. is correct

1.093  Most abundant subunit protein in actin filaments.

    B. is correct

1.094  This regulatory protein is not present in smooth muscle.

    G. is correct

1.095  Responsible for breaking crossbridges formed between thick and thin filaments in skeletal muscle.

    J. is correct

1.096  A protein that is not a component of thin filaments found in skeletal muscle.

    D. is correct

**Use the following to answer questions 97-100.**

    A.  protein
    B.  chloride
    C.  sodium
    D.  calcium
    E.  potassium
    F.  hydrogen
    G.  water

1.097  Increase in body content causes an increase in the volume of extracellular fluid.

C. is correct

1.098  A diffusible anion whose concentration in interstitial fluid is slightly greater than in plasma due to the Donnan effect.

B. is correct.

1.099  Unequal distribution across capillary endothelia results in the Donnan effect.

A. is correct

1.100  Principal cation in gastric juice.

F. is correct

# SECTION 2:    CARDIOVASCULAR

2.001  On an electrocardiogram ventricular depolarization is represented by:
    A.  the P wave
    B.  the QRS complex
    C.  the P-R interval
    D.  the A-V node
    E.  the S-T interval

B. is correct.
The waves on the electrocardiogram may be interpreted as follows: P = atrial depolarization, QRS - ventricular depolarization, T- ventricular repolarization.

2.002  The crest of the "A" wave in atrial pressure recordings marks the precise time when:
    A.  atrial contraction occurs
    B.  ventricular contraction occurs
    C.  the aortic valve closes
    D.  the mitral valve opens
    E.  the "Q" wave of the ECG is inscribed

A. is correct
The A-wave in atrial pressure readings corresponds to atrial systole (ventricular filling). It is followed by the C-wave, due to A-V valves "backward" bulging due to rising ventricular pressure, and then the V-wave, which corresponds to atrial filling.

2.003  Human coagulation proteins are activated to serine proteases during the coagulation process. One which is not activated but acts as a cofactor is:
    A.  II
    B.  X
    C.  VIII
    D.  XI
    E.  XII

C. is correct
Factor VIII, also called antihemophilic factor, is an essential cofactor in both the intrinsic and extrinsic clotting pathways. It is the factor which is missing in people with classic hemophilia.

2.004  The normal human heart rate may be varied in several ways. An increased heart rate results from stimulation of the:
    A.  vagus nerve
    B.  parasympathetic nerves
    C.  sympathetic nerves
    D.  adrenal cortex
    E.  carotid pressoreceptors

C. is correct
The pacemaker of the heart, the SA node, is under the influence of the autonomic nervous system. The sympathetic nervous system will increase its firing thereby increasing heart rate. The opposite is true of the parasympathetic nervous system.

2.005  The frequency of impulse discharge by the carotid sinus baroreceptors (pressure receptors) is decreased by a(n):
    A.  decrease in arterial pH
    B.  decrease in arterial blood pressure
    C.  increase in arterial blood pressure
    D.  decrease in arterial $O_2$ tension
    E.  increase in cardiac output

B. is correct
The carotid sinus baroreceptors are pressure receptors which are sensitive to stretch of the carotid sinus caused by increased blood flow through the sinus.

2.006  Ectopic ventricular stimulation is ineffective if applied during the:
    A.  atrial contraction
    B.  ventricular contraction
    C.  ventricular relaxation
    D.  entire cardiac cycle
    E.  P wave

B. is correct
Ventricular stimulation during contraction would in effect be during the refractory period, and thus would be ineffective.

2.007 The difference between the systolic and diastolic pressure is called:
- A. mean arterial pressure
- B. venous pressure
- C. circulation pressure
- D. pulse pressure
- E. mean pressure

D. is correct
By definition, pulse pressure is the difference between systolic and diastolic.

2.008 The best expression of the functional ability of ventricles is depicted by plotting ventricular output versus:
- A. stroke volume
- B. stroke work
- C. peripheral resistance
- D. right atrial pressure
- E. venous return

D. is correct
Venous return can be thought of as the "preload" of ventricle- the "input"; and compared with the "output" as an expression of ventricular functional ability.

2.009 At a heart rate of 72 beats per minute, the longest electrocardiogram interval would generally be:
A. duration of the P wave
B. duration of the QRS complex
C. P-R interval
D. Q-T interval
E. duration of the T wave

D. is correct
In the normal heart, the longest part of cardiac cycle is ventricular depolarization and repolarization; i.e. the Q-T interval.

2.010 Excitability is one of the three cardinal features of cardiac tissue, and varies in different parts of the cardiac cycle. All of the following statements are true, EXCEPT:
- A. the absolute refractory period corresponds to the QRS period recorded in the ECG
- B. the relative refractory period corresponds to the ST and T wave period in the ECG
- C. the relative refractory period is increased as the heart rate increases
- D. during a short period of the refractory period the heart is prone to fibrilation even with low intensity stimuli
- E. there is a supranormal state of excitability shortly after the end of the refractory period

C. is correct
The relative refractory period is a characteristic of the tissue's ability to move ions across its membrane, and remains constant irrespective of heart rate.

2.011 During the isometric contraction phase of the cardiac cycle, in the absence of abnormality, ventricular volume:
- A. decreases
- B. increases
- C. remains unchanged
- D. decreases as the intraventicular pressure increases
- E. increases as the intraventricular pressure increases

C. is correct
Isometric contraction, also called isovolumetric contraction, as it's name implies is a contraction in which volume is held constant. Its purpose is to increase intraventricular pressures above that of aortic pressures.

**Use the following data to answer question 12.**

| | |
|---|---|
| Oxygen consumption | 185 ml/min |
| Oxygen tension femoral artery | 20 volumes % (= 100 mmHg) |
| Oxygen tension femoral vein | 13 volumes % not mixed (= 35 mmHg) venous blood |
| Oxygen tension pulmonary artery | 14 volumes % (= 40 mmHg) |

2.012  Given the above data on a patient, the cardiac output would be, closest to:
A. less than 20 dl/min
B. between 20 and 24 dl/min
C. between 24 and 28 dl/min
D. between 28 and 32 dl/min
E. more than 32 dl/min

D. is correct

$O_2$ consumption

$$\frac{}{[A_{O_2}] - [V_{O_2}]}$$

$$\frac{185 \text{ ml/mm}}{6} = 30 \text{ dl/min}$$

2.013  The average output of the right and left heart must be maintained equally with great precision over long periods of time, or blood will accumulate in the systemic or pulmonary circulation. The most important factor in maintaining the two outputs dependent upon the length to which it is equal is:
  A. the length-tension relationship of cardiac muscle (Starling's law)
  B. sympathetic nervous outflow to the heart
  C. parasympathetic nervous outflow to the heart
  D. the baroreceptor system
  E. humoral sympathomimetic agents

A. is correct
Starling's Law states that the tension develops when cardiac muscle tissue is stretched. In order for the inputs of right and left heart to match, their inputs (of stretch) must be similar; i.e. the preloads must be closely related.

**Use the following data to answer question 14.**

| | |
|---|---|
| Mean capillary hydrostatic pressure | +30 mmHg |
| Plasma oncotic pressure | -28 mmHg |
| Tissue hydrostatic pressure | -5 mmHg |
| Tissue oncotic pressure | +8 mmHg |

2.014  Given the above data, there will be a net force (in mmHg) tending to move water from capillary to tissue of:
  A. 71
  B. 45
  C. 35
  D. 25
  E. 5

E. is correct
Increased hydrostatic pressure serves to force $H_2O$ towards an area of lower hydrostatic pressure, where oncotic pressures draw $H_2O$ into higher pore areas from lower pressure areas. Therefore, in the above example the net hydrostatic pressure is +25 (the designates movement from cap. to tissue) and the net oncotic pressure is -20. ∴ total net P = +5.

2.015  In atrial fibrillation, the ventricular rate is a function of the:
   A. rate of diastolic depolarization of the sinoatrial node
   B. conduction speed during endocardial to epicardial ventricular activation
   C. conduction speed through the bundle of His
   D. conduction speed through the ventricular septum
   E. refractory period of the atrioventricular junction

E. is correct
In atrial fibrillation conduction impulses are being sent (towards the ventricles) much further than normal. Thus, the rate of contraction of the ventricle must be a function of the ability of the conduction system to conduct the impulse to it; that is when it is not refractory.

2.016  In a normal person at rest at room temperature (22 °C), the organ tissue with the lowest arteriovenous oxygen difference is the:
   A. brain
   B. heart
   C. kidney
   D. liver
   E. skeletal muscle

E. is correct
Skeletal muscle at rest does not require a large metabolic demand relative to the other tissues and thus is the best choice.

2.017  Myocardial oxygen consumption is most closely correlated with:
   A. heart rate
   B. systolic pressure x heart rate
   C. stroke volume x heart rate
   D. stroke volume
   E. pulse pressure

B. is correct
$O_2$ consumption is approximately equal to work; work = rate + pressure + heat pressure; $\approx$ systolic pressure. Thus, f(rate) + f (sys. press.) equals myocardial oxygen consumption.

2.018  In a given individual, the pulse pressure is most closely related to the:
   A. stroke volume, assuming a constant cardiac output
   B. heart rate, assuming a constant stroke volume
   C. total peripheral resistance, assuming a constant arterial pressure
   D. cardiac output, assuming a constant stroke volume
   E. mean arterial pressure, assuming a constant peripheral resistance

A is correct
Remember pulse pressure is the difference between systolic and diastolic pressure. Greater stroke volume = higher systolic pressure - rate = constant, cardiac output therefore, diastolic pressure changes minimally.

2.019  Electrical stimulation of the distal end of a cut vagus nerve in the neck would be expected to:
   A. produce severe pain sensations even during general anesthesia
   B. increase the interval between successive R waves in the electrocardiogram
   C. decrease the P-R interval in the electrocardiogram
   D. increase the frequency of afferent impulses from the baroreceptors of the carotid sinus
   E. cause an immediate increase in cerebral blood flow

B. is correct
Vagal stimulation releases acetylcholine (ACh) on the SA node (pacemaker) of the heart, decreasing it's firing f, effectively decreasing heart rate. With a decreased heart rate, the time between successive "beats" (lub-dubs) would have to increase.

2.020 Normal ventricular depolarization:
  A. begins on the epicardial surface
  B. begins on the right side of the intraventricular septum
  C. begins on the left side of the intraventricular septum
  D. occurs during atrial contraction
  E. depends on syncytial connections between atrial and ventricular muscle fibers

C. is correct
Normal ventricular depolarizations begin on the left side of the intraventricular septum due to the anatomical composition of the left bundle branch.

2.021 Which of the following is most likely to occur as a result of strenuous exercise?
  A. a decreased volume of the thorax
  B. a rise in mean arterial blood pressure immediately upon an increased circulation time
  C. an increased circulation time
  D. a decreased number of impulses traveling in the carotid sinus nerve
  E. an increased arteriovenous oxygen difference

E. is correct
Increased $O_2$ demands by skeletal muscles during strenuous exercise, as well as increased amounts of metabolites (lactate, $CO_2$, etc.) would certainly cause an increase in the A-V $O_2$ difference.

2.022 The compensatory cardiovascular adjustments to assumption of the erect posture result in each of the following, EXCEPT:
  A. increased vasoconstriction in the distribution of the dorsalis pedis artery
  B. decreased renal blood flow
  C. decreased cerebral blood flow
  D. increased tonus of leg veins
  E. increased total peripheral resistance

C. is correct
Assuming a standing position elevates arterial pressure in the legs and feet and increases total peripheral resistance. Cerebral blood flow, however, is held constant.

2.023 During ventricular diastole in a supine individual, pressure is lowest in the:
  A. inferior vena cava
  B. right atrium
  C. right ventricle
  D. pulmonary artery
  E. aorta

C. is correct
During ventricular diastole, the ventricles are "waiting" to be filled by atrial blood. Since blood flows from areas of increased P to decreased P, then the pressure drop would be: inferior vena cava → right atrium → right ventricle (the lowest).

2.024 All of the following are compensatory adjustments to acute blood loss, EXCEPT:
  A. reflex increase of cardiac contractility
  B. vasoconstriction in splanchnic and somatic arterioles
  C. mobilization of extracellular fluid
  D. diminished rate of secretion of antidiuretic hormone
  E. reduction of the volume of blood contained in the pulmonary capillaries

D. is correct
ADH works to increase $H_2O$ reabsorption from the kidney, and therefore increase blood volume, which is a primary objective in acute blood loss.

2.025  A falsely high systolic arterial blood pressure is likely to be obtained using:
    A.  an extra width cuff on an infant's arm
    B.  an extra width cuff on an obese arm
    C.  a normal width cuff on an infant's arm
    D.  a normal width cuff on an obese arm
    E.  a normal width cuff on the forearm of a normal weight adult

D. is correct
An obese arm placed into a cuff of normal size will result in a falsely high reading because excess tissue must be compressed (increasing cuff pressure) to achieve brachial artery constriction.

2.026  During severe muscular exercise, the ability of the heart to meet increased metabolic demands is mainly dependent upon:
    A.  increased oxygen extraction from coronary arterial blood
    B.  increased oxygen supply through markedly increased coronary blood flow
    C.  increased diastolic volume of the ventricles
    D.  decreased oxygen tension in the coronary sinus
    E.  elevation of arterial $PCO_2$

B. is correct
The ability of the heart to perform it's duties as a muscular "pump' under conditions of severe exercise is dependent upon the availability of nutrients and oxygen supplied by adequate blood flow.

2.027  In a normal man, cardiac output:
    A.  immediately increases when a person is moved from a horizontal to an upright position on a tilt table
    B.  is the amount of blood pumped by each ventricle per beat
    C.  increases during exercise because of an increased peripheral resistance
    D.  increases when the filling pressure increases
    E.  increases whenever the aortic blood pressure increases

D. is correct
Starling's Law - the energy of contraction is proportional to the initial length of the cardiac muscle fiber. Thus, the extent of the preload is proportionate to the end diastolic volume.

2.028  Two capillary beds, I and II, are in parallel. 10 ml/second of blood flows through I, and 20 ml/second of blood flows through II. There is a gradient in blood pressure of 90 mmHg across I and II from arteries to veins. Total vascular resistance of the parallel system would be:
    A.  $3 \dfrac{mmHg}{ml/second}$

    B.  $3.33 \dfrac{ml/second}{mmHg}$

    C.  $30 \dfrac{mmHg}{ml/second}$

    D.  $0.33 \dfrac{mmHg}{ml/second}$

    E. none of the above

A. is correct
$WP = QR$
$R_T = 3$

$$R_1 = \frac{WP}{Q} = \frac{90}{10} = 9 PRU$$

$$R_2 = \frac{90}{20} = 4.5\ PRU$$

$$\frac{1}{R_T} = \frac{1}{R_1} + \frac{1}{R_2} = \frac{1}{9} + \frac{1}{4.5} = \frac{3}{9}$$

2.029 Normally, the largest fraction of the blood volume is found in:
  A. venous vessels
  B. arterial vessels
  C. capillaries
  D. heart chambers
  E. none of the above

A. is correct
The venous system is known as the "reservoir" of the blood in that it normally houses the largest amount of total blood volume.

2.030 The major cause of the increase in skeletal-muscle blood flow during vigorous exercise is:
  A. increase in skeletal-muscle metabolism
  B. neurogenic cholinergic dilatation
  C. epinephrine
  D. increased arterial pressure
  E. the skeletal-muscle pump

A. is correct
Blood flow to skeletal muscle is mainly dependent upon "local effects"; primarily being the metabolic and nutritional demands of the muscle tissue. With exercise, the skeletal-muscle blood flow would increase.

2.031 In man, simultaneous recording of electrocardiographic patterns, heart sounds, and carotid pulse will disclose simultaneity of:
  A. the QRS complex, and the first heart sound
  B. the second heart sound, and the T wave
  C. the P wave and onset of the first heart sound
  D. the peak of the QRS complex, and onset of rise in carotid pressure
  E. the peak of the QRS complex and the beginning of isovolumetric contraction

E. is correct
Excitation precedes contraction.

2.032 The electrocardiographic pattern PQRST-P-PQRST-P-PQRST-P-PQRST indicates:
  A. atrial premature systoles
  B. atrial fibrillation
  C. 2:1 atrioventricular block
  D. ventricular tachycardia
  E. bundle-branch block

C. is correct
AV blocks occur at the locus between the atria and ventricles: the AV node. The "block" is a situation in which the depolarization through the atria is not allowed to continue to the ventricles, thus giving the characteristic ECG in which there is one (1) block for every two (2) "attempts" at the full cardiac cycle; i.e. 2:1 block.

2.033 As a result of significant elevation in intrapericardial pressure:
  A. systemic blood pressure rises above control because of increased peripheral resistance
  B. systemic blood pressure falls below control because of decreased peripheral resistance
  C. systemic blood pressure remains unchanged because of appropriate compensation
  D. systemic blood pressure falls because of decreased stroke volume
  E. pulmonary arterial pressure falls but systemic arterial pressure rises

D. is correct
With significant elevation of intrapericardial pressure diastole is impeded; End diastolic volume is decreased and stroke volume is decreased.

2.034  Most vascular smooth muscle is controlled by:
    A.  the thoracolumbar sympathetic portion of the spinal cord
    B.  cranial nerves
    C.  sacral outflows from the spinal cord
    D.  somatic motor ($\beta$) fibers
    E.  gamma afferent fibers

A. is correct
Total peripheral resistance is controlled via the sympathetic nervous system.

2.035  In a vascular bed, a rise in peripheral resistance flow is indicated if:
    A.  flow and mean arterial pressure are unchanged; venous pressure rises
    B.  flow and mean arterial pressure are decreased proportionally; venous pressure is unchanged
    C.  flow and venous pressure increase; mean arterial pressure is unchanged
    D.  mean arterial pressure and venous pressure are unchanged; flow increases
    E.  mean arterial pressure increases; flow and venous pressure are unchanged

E. is correct
By Ohm's Law: ( pressure = flow x resistance) - (P = Q $\Sigma$ R - an increase in peripheral resistance to flow must be accompanied by an increase in arterial pressure (the pressure before the resistance) if flow is to remain the same. An example of this situation in vivo could be contraction of a pre-capillary arteriole.

2.036  Which of the following stimulates the production of erythrocytes?
    A.  nervous stimulation of bone marrow
    B.  administration of epinephrine
    C.  administration of levarterenol
    D.  generalized tissue hypoxia
    E.  administration of cortisone

D. is correct
The main stimulus for erythrocyte formation is the hormone erythropoeitin which is produced and released by the kidney in response to hypoxia.

2.037  The part of the mammalian heart with the slowest rate of impulse conduction is the:
    A.  sinoatrial node
    B.  atrial muscle
    C.  atrioventricular node
    D.  bundle of His
    E.  ventricular muscle

C. is correct
There is a slowing of the impulse conduction through the junctional fibers and AV node, which allows time for ventricular filling prior to contraction.

2.038  The pulmonary vascular system:
    A.  has a higher volume flow of blood than the systemic vascular system
    B.  contains less blood when the person is in a vertical position than when he is in a horizontal position
    C.  exhibits a high degree of sympathetic tone
    D.  is a low pressure, high resistance system
    E.  normally contains approximately 50% of the total blood volume

B. is correct
In the upright postion, the upper portions of the lungs are above the level of the heart and the bases are below it. Therefore, there is a linear decrease in pulmonary blood flow from the apices to the bases of the lungs.

2.039  Cerebral blood flow increases markedly;
    A.  under reflex vasomotor control of cerebral vessels
    B.  with declining concentrations of $CO_2$ in the blood
    C.  with increasing arterial $PCO_2$
    D.  with increasing arterial $PO_2$
    E.  during elevated intracranial (extravascular) pressures

C. is correct
Increasing arterial $PCO_2$ leads to stimulation of the chemosensitive area of the medullary respiratory center by dissociation of $H_2CO_3$ (from $CO_2$ and $H_2O$) forming $H^+$ ions. CNS hypoxia, respiratory drive and cerebral blood flow are increased.

2.040 The frequency of firing of carotid sinus baroreceptors is increased by:
A. decreased mean arterial pressure
B. reduced blood $O_2$ content
C. reduced blood $CO_2$ content
D. increased pulse pressure
E. none of the above

D. is correct
The carotid baroreceptors are mechanoreceptors which are sensitive to changes in pressure: their firing rate increases with increased pressure, and decreases with decreased pressure.

2.041 An unconscious patient, badly bruised and beaten all over his trunk and legs, is dropped at the door of a well-stocked hospital by an unknown criminal who speeds away without leaving any information. The patient is pale, cold and clammy with blood pressure 70/50 mmHg, heart rate 120 beats/minute, hematocrit 55%. From the evidence cited, the doctor should:
A. keep the patient propped up in bed to avoid pulmonary edema
B. infuse isotonic saline intravenously to combat dehydration
C. infuse plasma to restore plasma volume
D. infuse whole blood
E. administer drugs to increase heart rate

C. is correct
In severe shock, the coronary blood flow is reduced because of hypotension and tachycardia. The physician should improve the reduction of myocardial function by re-expansion of the blood volume.

2.042 Compared with a sedentary individual, a trained athlete has all of the following, EXCEPT:
A. a lower resting heart rate
B. greater stroke volume
C. a larger rise in blood lactic acid for a comparable work load
D. a smaller rise in pulse rate for a comparable work load
E. greater cardiac reserve

C. is correct
The trained athlete would be expected to "handle" the same work load as the sedentary person more aerobically and thus not produce as much lactic acid.

2.043 Knowledge that a drug causes a generalized dilation of both pre- and postcapillary resistance vessels permits the conclusion that the drug will cause:
A. an elevation of capillary filtration pressure
B. a reduction in capillary hydrostatic pressure
C. a decrease in capillary hydrostatic pressure
D. an increase in vascular capacity
E. a decrease in plasma colloidal osmotic pressure

D. is correct
A relaxation of pre- and postcapillary sphincter will increase the number of capillaries that are dilated and will result in an increase of vascular capacity.

2.044 A reflex fall in mean arterial blood pressure occurs as a result of:
A. decrease in cardiac output
B. reduction of renal blood flow by renal vasoconstriction
C. increase of mean blood pressure in the aortic arch
D. decrease of pulmonary blood flow as a result of decreased venous return
E. reduction in amplitude of arterial pressure pulsation at constant mean pressure

C. is correct
Increased blood pressure in the aortic arch will result in a reflex bradycardia and fall in arterial blood pressure.

2.045 Mr. A is consuming oxygen at the rate of 200 ml/minute, Mr. B is consuming oxygen at the rate of 2,000 ml/minute. Both are adult men of normal size. The most probable single explanation for Mr. B's relatively high rate of oxygen consumption is that he:
  A. has greater thyroid activity
  B. is exercising vigorously
  C. has a high fever
  D. is very excited emotionally
  E. has just eaten a big steak dinner

B. is correct
The most probable explanation for an increased oxygen consumption is that it is due to exercise.

2.046 The sinoatrial node is the normal pacemaker because it:
  A. is most richly supplied with nerve endings
  B. is specially modified, "nodal" myocardium
  C. is the first portion of the heart to beat in the embryo
  D. possesses the highest frequency of automatic discharge
  E. is the origin of the conduction system of the heart

D. is correct
Because the SA node has the highest rate of discharge, its action potentials are completed prior to all others, and thus sets the pace of the entire system.

2.047 The venous return to the heart increases transiently when:
  A. a sudden increase in blood volume occurs
  B. the abdomen is suddenly compressed
  C. the sympathetics are suddenly stimulated
  D. the arterioles suddenly dilate
  E. all of the above

E. is correct
All the factors will cause a transient increase in venous return.

2.048 Suppression of activity in the sinoatrial node of a normal heart would be likely to result in:
  A. assumption of pacemaker activity in the artrioventricular node
  B. shortening of the Q-T interval
  C. inversion of the P waves
  D. prolongation of the P-R interval
  E. A and C

E. is correct
When SA node activity is suppresed an alternative lower pacemaker, the AV node, takes over to maintain blood pressure. Because the AV node discharges irregularly, the ventricles beat irregularly where the P wave is usually buried in the QRS.

2.049 If blood flow through the brain is reduced to levels inadequate to sustain consciousness, changes in cardiovascular dynamics would include:
  A. pulmonary congestion
  B. decrease in renal vascular resistance
  C. elevated systemic pressure accompanied by increased rate of arterial run-off and decreased rate of venous return
  D. intense systemic vasoconstriction with elevated blood pressure
  E. increased splanchnic blood flow

D. is correct
Fainting or syncope is most commonly due to peripheral vascular or cardiac abnormalities that cause inadequate cerebral blood flow. The hypotension is antagonized by increased systemic vasoconstriction to increase blood pressure.

2.050  Variables which increase during exercise include:
    A.  venous return
    B.  total peripheral resistance
    C.  pulse pressure
    D.  duration of systolic ejection
    E.  A and C above

E. is correct
Exercise increases venous return by the increased activity of the respiratory pump and the muscle pump.

2.051  In coronary circulatory function:
    A.  arterial inflow increases during isovolumetric contraction
    B.  venous outflow decreases during systole
    C.  neurogenic vascular control is normally the principal determinant of resistance to flow
    D.  resistance decreases when myocardial oxygen consumption increases
    E.  $b_1$ receptors control resistance to flow

D. is correct
One would expect that increasing demand of nutrients by the myocardium would cause local vasodilatory effects in the coronary circulation, and thus decrease resistance to blood flow.

2.052  During muscular exercise the cardiac output is augmented by:
    A.  the kneading of adjacent veins by active muscles
    B.  increased intrathoracic pressure
    C.  increased cholinergic discharge to the heart
    D.  increased arterial pressure
    E.  decreased adrenergic discharge to the heart

A. is correct
The squeezing of veins in exercise will increase venous return and therefore increase cardiac output.

2.053  An experimental animal is given a drug that causes a 50% increase in myocardial oxygen consumption. Likely sites of action of the drug would include the:
    A.  coronary vessels
    B.  sinoatrial node
    C.  atrioventricular node
    D.  peripheral arterioles
    E.  all of the above

E. is correct
A drug which would increase myocardial $O_2$ consumption would likely affect heart rate (SA and AV nodes), total peripheral resistance (peripheral arterioles), and coronary blood flow.

2.054  Stimulating the parasympathetic nerves to the heart results, directly or indirectly, in:
    A.  an increased heart rate
    B.  a diminished conduction velocity through the atrioventricular node
    C.  a decreased permeability of the sinoatrial node
    D.  vasoconstriction of the coronary vessels
    E.  increased TPR

B. is correct
The parasympathetic nerves to the heart (the vagi) release ACh into the SA node and the AV junctional fibers. ACh causes a decrease in excitability of the junctional fibers thereby slowing the conduction velocity through the AV node.

2.055  The epicardial surface of ventricular muscle:
    A.  depolarizes before the endocardial surface of the ventricle
    B.  repolarizes after the endocardial surface
    C.  depolarizes after the endocardial surface of the ventricle
    D.  has the slowest conduction velocity
    E.  B and C

C. is correct
The depolarization wave spreads along the ventricular walls to the AV groove proceeding from the endocardial to the epicardial surface.

2.056 Upon echocardiography a 3-month-old child is diagnosed with having the most common congenital heart defect. At this time cardiac auscultation would reveal:
   A. a diastolic murmur
   B. an early systolic murmur
   C. a pansystolic murmur
   D. a "machinery murmur"
   E. a "functional murmur"

C. is correct

The most common heart defect is a ventricular septal defect which causes a shunt of blood from the right to left ventricle during systole. Thus, a sound is generated throughout systole and it is therefore known as a pan (whole) systolic murmur.

2.057 An increase in afterload is better tolerated by the left rather than the right ventricle. This may be explained by the fact that:
   A. the left ventricle is thicker than the right
   B. the left ventricle is better anatomically situated
   C. the left ventricle has a richer sympathetic nerve supply
   D. the left ventricle has a greater blood supply
   E. the left ventricle receives a more abundant vagal innervation

A. is correct

The greater force of ejection by the left ventricle is achieved in part by the greater muscle mass of the left ventricle. This larger ventricular mass allows the left ventricle to tolerate an afterload more efficiently.

2.058 By increasing cardiac contractility:
   A. stroke volume is unchanged at a higher left ventricular end diastolic volume
   B. left ventricular end diastolic volume will be increased with the reduction in the maximum velocity of shortening
   C. an increase in stroke volume can occur if the initial fiber resting length remains the same
   D. stroke volume can remain the same in spite of an increase in left ventricular and end diastolic pressure
   E. both C and D

C. is correct

Stroke volume = end diastolic volume - end systolic volume.

2.059 The most likely action of a drug that prolongs the P-R interval of the electrocardiogram is:
   A. a reduced conduction velocity of the atrial fibers
   B. a reduced conduction velocity of the ventricular fibers
   C. a hyperpolarization of the atrioventricular nodal fibers
   D. an increased conduction velocity of the atrial fibers
   E. an increased force of contraction of the right ventricle

A. is correct

With a prolonged P-R interval, the conduction between the atria and ventricles is slowed. For example, in first degree heart block all atrial impulses reach the ventricles but the P-R interval is abnormally long.

2.060 The direct result of calcium influx, an action potential, is:
   A. an augmentation of depolarization
   B. an acceleration in the rate of the neurotransmitter production
   C. an increase in permeability of the presynaptic membrane
   D. fusion of the neurotransmitter containing vesicles to the presynaptic membrane
   E. hyperpolarization of the presynaptic membrane

D. is correct

It is believed that the influx of calcium into the presynaptic membrane is responsible for the fusion of the neurotransmitter containing vesicles to the presynaptic membrane so that they maybe extruded into the synaptic cleft.

2.061 Which of the following is most likely to occur in a healthy subject in response to strenuous running?
  A. a decrease in the ejection fraction of the ventricles
  B. a decrease in the maximum dP/dt of the left ventricle
  C. an increase in circulation time
  D. a decrease in arteriovenous $O_2$ difference
  E. an increase in concentration of arterial lactate

E. is correct
During exercise there is an increased blood flow to the exercising muscle and consequently an accumulation of vasodilator metabolites such as lactic acid.

2.062 In a healthy subject, running is usually associated with a decrease in the end-systolic volume of the right and left ventricles and with an increase in their stroke volume. The mechanism for this response is probably the following:
  A. Starling's law of the heart
  B. an increase in sympathetic tone to the ventricles
  C. an increase in parasympathetic tone to the ventricles
  D. a decrease in venous return of blood to the heart
  E. an increase in pulmonary and system resistance

B. is correct
During muscular exercise there is an increased sympathetic discharge so that myocardial contractility and heart rate are increased.

2.063 Each of the following is an important mechanism for increasing the blood flow at active skeletal muscle during exercise, EXCEPT:
  A. an increased cardiac output
  B. an increased concentration of epinephrine in the vicinity of the blood vessels of active skeletal muscle
  C. an increased concentration of norepinephrine in the vicinity of the blood vessels of active skeletal muscle
  D. an increased concentration of acetylcholine in the vicinity of the blood vessels of active skeletal muscle
  E. the action of metabolites on skeletal muscle blood vessels

C. is correct
Norepinephine will cause vasoconstriction of the blood vessels of active skeletal muscle and therefore decrease blood flow.

2.064 The Bainbridge reflex refers to:
  A. the chemoreceptor reflex
  B. the aortic baroreceptor reflex
  C. the carotid sinus reflex
  D. the central venous pressure reflex
  E. the sympathetic respiratory reflex

D. is correct
Rapid infusion of blood in anesthetized animals can produce an increase in heart rate if it was initially low. This is known as the Bainbridge reflex described by Bainbridge in 1915. The reflex competes with the baroreceptor-mediated decrease in heart rate.

2.065 Sino-atrial myocardial pacemaker fibers are distinct from myocardial fibers in that during the time lapse between repetitive action potentials the sodium conductance seems to be uniformly high while the potassium conductance is:
  A. continuously increasing from a low level
  B. continuously decreasing from a high level
  C. undergoing no change at all
  D. going through a biphasic change, high first, low later, and finally high again at the end of the electrical diastole

B. is correct
Nodal tissue is characterized by an action potential with a low amplitude and a low conduction velocity. In phase 4, sodium conductance starts low and steadily increases while potassium outward movement starts high and steadily decreases.

2.066  A greater pulse pressure is positively correlated with:
   A.  large systolic stroke volume
   B.  higher compliance of the arterial walls
   C.  higher frequency of contraction of the heart, with a constant cardiac output
   D.  reduced period of systolic ejection
   E.  reduced venous return

A. is correct
Increased stroke volume

2.067  A 73-year-old patient on the cardiac service suffers from an arrhythmia which prevents the electrical impulse from slowing down at the AV node. Which of the following would be a direct result of this type of arrhythmia?
   A.  insufficient time for ventricular filling
   B.  augmentation of ventricular contraction
   C.  increase in venous return
   D.  insufficient time for atrial filling
   E.  increased aortic resistance

D. is correct
Under normal conditions when the electrical impulse reaches the A-V node, it slows down so that the atria may fill with blood before atrial contraction. If this slowing is prevented the atria would contract prior to their filling and less blood would reach the ventricles.

2.068  The coronary arteries actively supply blood to the heart during which of the following periods of the cardiac cycle?
   A.  between $S_1$ and $S_2$
   B.  during the period of isovolumetric contraction
   C.  during $S_3$
   D.  between $S_2$ and $S_1$
   E.  between $S_2$ and $S_3$

D. is correct
Coronary blood flow to the subendocardial region of the left ventricle occurs only during diastole. There is no coronary blood flow to this region during systole. Diastole begins after the closure of the aortic valve ($S_2$) and ends with the closure of the A-V valve ($S_1$)

2.069  If the atria were denervated approximately what percentage of blood would reach the ventricles?
   A.  10
   B.  30
   C.  50
   D.  80
   E.  100

D. is correct
As blood enters the atria approximately 80% passively flows into the ventricles. The atria are responsible for pumping the remaining 20% into the ventricles. If the atria are denervated, the atria will be unable to deliver the last 20%.

2.070  A 68-year-old white male suffers a myocardial infarction and in the process also infarcts a few of the papillary muscles of his mitral valve. Upon cardiac auscultation one would expect to hear:
   A.  a pansystolic murmur
   B.  an early systolic murmur
   C.  a diastolic rumble
   D.  a "machinery murmur"
   E.  normal heart sounds

B. is correct
With myocardial infarction, a new murmur of mitral regurgitation may be present. Since this regurgitation occurs during systole it is an early systolic murmur.

2.071 Aortic incompetency leads to:
A. high end-diastolic ventricular volume (EDVV)
   low end-diastolic aortic pressure (EDAP)
   low cardiac output (CO)
   large systolic ventricular stroke volume (SVSV)
B. low EDVV
   high EDAP
   high CO
   small SVSV
C. low EDVV
   high EDAP
   low CO
   large SVSV
D. high EDVV
   low EDAP
   high CO
   small SVSV
E. high EDVV
   high EDAP
   high CO
   large SVSV

A. is correct
Aortic incompetency causes a portion of the left ventricular stroke volume ejected during systole to regurgitate into the left ventricle during diastole. This produces an increase in end-diastolic volume with an increase in total stroke volume. Left ventricular forward output decreases and there is a reduction in systemic diastolic blood pressure.

2.072 The right ventricle may be expected to have a greater thickness than that of the left ventricle in cases of:
A. aortic stenosis
B. aortic insufficiency
C. patent ductus arteriosus
D. pulmonary artery stenosis
E. all of the above

D. is correct
With pulmonary artery stenosis there is chronic regurgitation into the right ventricle which stimulates sarcomere replication producing a thickening of the right ventricle.

**Use the following figure to answer questions 73 and 74**

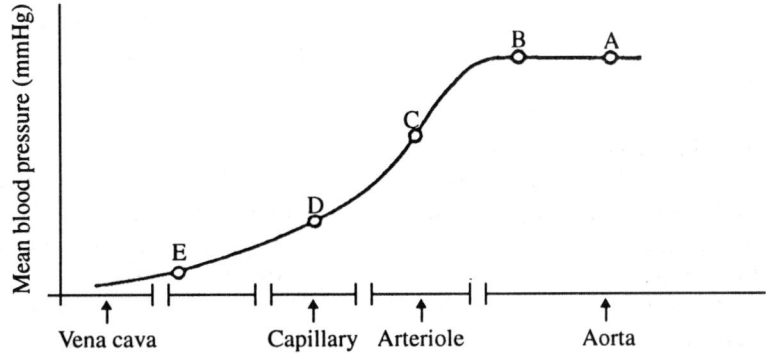

The figure above depicts the change in mean systemic blood pressure as the blood goes from the aorta through the various vessel types.

2.073 Indicate in which bed sites there occurs major vascular resistance

C. is correct
The greatest resistance to blood flow in the cardiovascular system is in the arterioles.

2.074 Indicate in which bed sites there occurs major fluid exchange

D. is correct
The capillaries are the major exchange vessels in the circulatory system.

2.075  Which of the following statements concerning cardiac excitation is correct?
- A.  the ventricular epicardial surface is depolarized before the endocardial surface
- B.  the ventricular epicardial surface is repolarized before the endocardial surface
- C.  the apex of the heart is the first part of the heart to become depolarized
- D.  with normal cardiac excitation pathways, the AV node is considered to be the cardiac pacemaker
- E.  the AV node is found in the interventricular septum

B. is correct
Depolarization of the ventricular muscle starts at the left side of the intraventricular septum then spirals down the septum to the apex of the heart. Depolarization returns along the ventricular walls from the endocardial to the epicardial surface. Repolarization occurs at the ventricular epicardial surface first then the endocardial surface.

2.076  The C wave of the jugular pulse occurs:
- A.  just prior to atrial systole
- B.  just after the peak of auricular systole
- C.  during ventricular systole
- D.  just after the dicrotic notch in aortic pressure
- E.  at the onset of expiration

C. is correct
The C wave is produced by the bulging of the tricuspid valve into the atria during isometric ventricular contraction.

2.077  The P wave of the ECG occurs at:
- A.  the beginning of atrial contraction
- B.  the end of atrial contraction
- C.  the beginning of ventricular contraction
- D.  the end of ventricular contraction
- E.  the time of the second heart sound

A. is correct
The P wave is produced by atrial depolarization.

2.078  An increase of left ventricular and diastolic pressure will be associated with a rise in left atrial and pulmonary capillary pressure, as left ventricle, left atrium and pulmonary veins communicate freely during ventricular diastole. Such an increase in pressure to a level equal to 15-30 mmHg is indicative of:
- A.  pulmonary artery hypertension
- B.  right ventricular failure
- C.  left ventricular failure
- D.  increased compliance of the pulmonary artery
- E.  coarctation of the aorta

C. is correct
Left ventricular failure is associated with increased pulmonary venous pressure and pulmonary venous distension leading to pulmonary congestion.

2.079  Extensive occlusion of pulmonary vessels by thrombi originating in the pelvis or in the lower extremities may result in:
- A.  left ventricular failure
- B.  right ventricular failure
- C.  increased cardiac output
- D.  complete compensation by excessive distensibility of non-occluded vessels

B. is correct
In pulmonary embolism, a thrombus arises elsewhere and migrates to the pulmonary vascular tree where it causes an obstruction. Obstruction of pulmonary arteries increases resistance to blood flow through the pulmonary circuit and increases right ventricular afterload.

2.080 If 5 milligrams of T-1824 dye are injected at time zero in a peripheral vein, and the first dye appears in an arm artery 15 seconds later, reaching an average concentration equal to 0.005 mg/ml, and essentially disappearing (by extrapolation) from circulation 20 seconds after the injection, we can calculate the cardiac output to be equal to:
- A. 1,000 ml blood/minute
- B. 2,000 ml blood/minute
- C. 3,000 ml blood/minute
- D. 4,000 ml blood/minute
- E. 5,000 ml blood/minute

D. is correct

Determination of cardiac output by indicator dilution.

$$Flow = \frac{Amt. \ dye \ injected}{Av. \ conc.}$$

$$\frac{5 \ mg}{0.005 \ mg/l} = 1 \ liter$$

Flow in 15 sec. = 1 liter in 15 sec.
Flow/min = 4 x 1000ml = 4,000 ml/min or 4 l/min

2.081 Increased pressure within the carotid sinus causes:
- A. an increase in heart rate
- B. a fall in venous pressure
- C. reflex bradycardia
- D. reflex hypernea
- E. reflex increase in venous pressure

C. is correct

Carotid sinus receptors monitor arterial circulation. Impulses generated in the baroreceptors execute the vagal innervation of the heart producing vasodilation and bradycardia.

2.082 Sympathetic vasoconstrictor tone is diminished in response to increased activity of:
- A. carotid sinus receptors
- B. medullary chemoreceptors
- C. pain receptors
- D. carotid body chemoreceptors
- E. respiratory activity

A. is correct

Impulses generated in the baroreceptors of the carotid sinus and aortic arch inhibit the tonic discharge of the vasoconstrictor nerves.

2.083 Cutting the carotid sinus nerves and the afferent vagal branches from the aortic arch baroreceptors will result in:
- A. a decrease in arterial pressure on head-down tilting
- B. partial or total loss of compensatory ability on head-up tilting
- C. a decrease in heart rate and arterial blood pressure
- D. a decrease in cerebral blood flow and venous pressure
- E. no change in control of blood pressure

B. is correct

Denervation of the carotid sinus results in an absence of compensatory mechanisms associated with positional changes.

2.084 All the following statements regarding Einthoven's limb lead II are correct EXCEPT:
- A. lead II, is measured between the left leg and the right arm
- B. lead II gives an upward deflection when the left leg is positive with respect to the right arm
- C. lead II gives an upward deflection to represent the normal wave for atrial depolarization
- D. lead II gives an upward deflection to represent the normal wave for ventricular repolarization
- E. lead II measures the potential difference between the right arm and the left arm

E. is correct

In lead II, the electrodes are on the right arm and left leg with the leg positive to the right arm.

2.085 The relatively long period of time (200 milli-seconds) during which the ventricular cardiac muscle membrane is depolarized (the plateau of ventricular action potentials) coincides with a certain portion of the peripherally obtained electrocardiogram. This is the:
  A. P wave
  B. P-Q interval
  C. U wave
  D. Q-T segment
  E. T wave

D. is correct
The Q-T segment represents the plateau (or phase 2) of the ventricular action potential.

2.086 Stimulation of the midline vasodilator area of the vasomotor center:
  A. inhibits the neighboring vasoconstrictor center
  B. sends fibers to the heart with the vagus nerve
  C. inhibits transmission through sympathetic ganglia
  D. inhibits impulses through the vagus nerve
  E. affects the aorta but not the other large arteries

A. is correct
The main control of blood pressure is executed by groups of neurons in the medulla known as the vasomotor center. Remember that impulses of the vasodilator area of the vasomotor center inhibit the vasoconstrictor center and execute the vagal innervation of the heart producing vasodilation and bradycardia.

2.087 In hemorrhagic shock, one would expect a decrease in all of the following, EXCEPT:
  A. cardiac work
  B. cardiac output
  C. coronary flow
  D. oxygen usage
  E. plasma lactate

E. is correct
In hemorrhagic shock, plasma lactate concentrations would not be altered since there is no association with exercise and increased metabolic work.

2.088 In congestive heart failure edema results from:
  A. an elevation of cardiac output
  B. an elevation of peripheral resistance
  C. an increase of plasma oncotic pressure
  D. an elevation of capillary hydrostatic pressure
  E. a decreased blood pressure

D. is correct
Cardiac edema occurs when the systemic hydrostatic venous pressure is greater than the systemic oncotic venous pressure. Cardiac edema occurs due to increased systemic venous pressure that results from right ventricular failure.

2.089 Essential hypertension is generally associated with an early increase in:
  A. cardiac work
  B. coronary flow
  C. myocardial oxygen consumption
  D. cardiac output
  E. renal filtration

A. is correct
Cardiac work through total oxygen consumption of the heart is increased in hypertension by the work of the heart expelling blood against an increased pressure.

**Use the following data to answer question 90.**

An adult male is seen in the clinic with the following findings:

| | |
|---|---|
| Cardiac output | = 5 l/min |
| Systolic pressure | = 200 mmHg |
| Diastolic pressure | = 120 mmHg |
| Heart rate | = 80 bpm |
| Plasma catecholamines | = normal |
| Blood volume | = 5 l |
| Total peripheral resistance | = increased |
| Plasma angiotension | = normal |

2.090. From the above data, it should be concluded that the patient is:

A. hypertensive due to excessive cardiac activity
B. in hypotension of renal origin
C. hypertensive as a result of excessive adrenomedullary secretion
D. suffering from essential hypertension
E. benign intracranial hypertension

**D. is correct**

In order for hypertension to occur there must be an elevation of cardiac output or total peripheral resistance or both. An abnormality in peripheral resistance is a major reason for most cases of hypertension. Hypertension that has no known cause is labeled primary or essential hypertension.

**Use the following to answer questions 91-95.**

Select the most appropriate cardiovascular parameter for each of the descriptions below. Each parameter may be used once, more than once or not at all.

A. cardiac output
B. diastolic pressure
C. end diastolic volume
D. end systolic volume
E. heart rate
F. length of time in diastole
G. length of time in systole
H. pulse pressure
I. stroke volume
J. systolic pressure
K. total peripheral resistance
L. V-central venous pressure work
M. A-V valve closure

2.091  During a normal, general physical examination stress test, this cardiovascular parameter will most likely increase by the greatest factor.

**E. is correct**

2.092  This cardiovascular parameter is the primary determinant of afterload.

**J. is correct**

2.093  With respect to a normal pressure/volume loop curve, this parameter will remain approximately constant, or fall slightly during exercise.

**B. is correct**

2.094  The end diastolic volume minus the end systolic volume.

**I. is correct**

2.095  Associated in time with the opening of the A-V valves.

**L. is correct**

**Use the following to answer questions 96-100.**

A. phase 2 of the ventricular myocardial action potentials
B. slow inward calcium current
C. P-Q interval of the electrocardiogram
D. ventricular repolarization
E. positive dipole approaching the exploring electrode
F. phase 3 of the myocardial action potential
G. rapid inward sodium current
H. S-T interval of the electrocardiogram
I. atrial repolarization
J. negative dipole approaching the exploring electrode
K. Einthoven's Law
L. rapid inward potassium current
M. first degree A-V heart block
N. second degree A-V heart block

2.096  The electrocardiographic pattern PQRST-P-PQRST-P-PQRST-P-PQRST?                          N. is correct

2.097  Diastolic depolarization of the sinoatrial nodal cell.                                                              B. is correct

2.098  Myocardial action potential within A-V nodal tissue.                                                              C. is correct

2.099  Positive ECG deflection.                              E. is correct

2.100  The sum of the deflections in leads I, II and III equals zero.                                                          K. is correct

# SECTION 3:   RESPIRATORY PHYSIOLOGY

3.001 The normal maximum voluntary ventilation (maximal breathing capacity) is approximately:
- A. 50-60 l/min
- B. 60-75 l/min
- C. 75-100 l/min
- D. 100-125 l/min
- E. 125-170 l/min

E. is correct
The normal range from maximal breathing capacity ranges from 125-170 l/min.

3.002 Regulation of breathing during early exercise is governed by:
- A. proprioceptor fibers alone
- B. motor cortex alone
- C. motor cortex and proprioceptor fibers
- D. mast cell liberation of histamine
- E. fall in partial pressure of carbon dioxide

C. is correct
Afferent nerve impulses from proprioceptors reflex increase respiration.

3.003 The lung volume at which small airway closure begins is referred to as the:
- A. expiratory reserve volume
- B. residual volume
- C. alveolar volume
- D. closing volume
- E. inspiratory reserve volume

D. is correct
The portion of the vital capacity remaining when airway closure begins is the closing volume.

3.004 At the end of expiration (no air flow) total alveolar gas pressure is:
- A. equal to pulmonary capillary pressure
- B. equal to intrapleural pressure
- C. less than at the end of inspiration (no air flow)
- D. equal to atmospheric pressure
- E. one-half the intrapleural pressure

D. is correct
During inspiration Alveolar pressure decreases to -1 cm of $H_2O$, and then returns to zero. During expiration alveolar pressure increases to +1 cm of $H_2O$ and then returns to zero at the end of expiration. Note: 0 cm $H_2O$ is equal to atm pressure.

3.005 The major amount of carbon dioxide carried by the blood is in:
- A. the dissolved state
- B. carbonic anhydrase
- C. bicarbonate ion
- D. carbaminohemoglobin
- E. combination with oxygen

C. is correct
$CO_2$ is transported in the blood mainly as bicarbonate. 5% dissolved in plasma, 5% as a carbamino compound, and 90% as bicarbonate ion.

3.006 The blood of an anemic individual with a hemoglobin concentration of 10 gm/100 ml blood breathing room air would have a(n):
- A. normal arterial $PO_2$ and normal arterial oxygen content
- B. reduced arterial $PO_2$ but a normal arterial oxygen content
- C. reduced arterial $PO_2$ and an arterial oxygen content of about 13 ml/100 ml blood
- D. normal arterial $PO_2$ but an arterial oxygen content of about 13 ml/100 ml blood
- E. arterial oxygen content of about 20 ml/100 ml

D. is correct
Each gram of hemoglobin can combine with 1.34 ml of $O_2$. In this anemic individual, therefore:

$$\frac{1.34 \text{ ml } O_2}{1.0 \text{ g Hb}} \text{ X} \frac{10g}{100 \text{ ml}} \quad \frac{13.4 \text{ ml } O_2}{100 \text{ ml blood}}$$

$PO_2$ is affected by dissolved $O_2$ and not by Hb concentration.

3.007 The following information is given for an individual: oxygen consumption 250 ml/min, arterial oxygen content 20 vols%, and mixed venous oxygen content 18 vols%. The cardiac output in l/min is:
A. 5.0
B. 7.5
C. 10.0
D. 12.5
E. 15.0

D. is correct
By the Fick Method:

$$CO = \frac{O_2 \text{ consumption ml/min}}{(A_{O_2}) - (V_{O_2})} =$$

$$CO = \frac{250 \text{ ml/min}}{200 \text{ ml/l} - 180 \text{ ml/l}}$$

$$\frac{250 \text{ ml/min}}{20 \text{ ml/l}} = 12.5 l/min$$

3.008 The pleura and pleural cavities are the sites of variety of medical and surgical disorders. At the end of inspiration the intrapleural pressure is normally:
A. minus 4 to minus 6 mmHg
B. minus 20 to minus 30 mmHg
C. atmospheric
D. plus 4 to plus 8 mmHg
E. equal to intrathoracic pressure

A. is correct
At the beginning of inspiration, intrapleural pressure is -2mmHg. At the end it falls to -6mmHg. At the end of expiration, pressure returns to -2mmHg.

3.009 If the atmospheric pressure is 760 mmHg and the percentage of nitrogen in the alveolar air is 79, the partial pressure of nitrogen would be approximately:
A. 40 mmHg
B. 46 mmHg
C. 100 mmHg
D. 280 mmHg
E. 600 mmHg

E. is correct
The composition of dry air is 21% $O_2$, 0.04% $CO_2$, 79% $N_2$ and 0.92% other inert constituents. The partial pressure of nitrogen is therefore: 760 mmHg x 0.79 or 600 mmHg.

3.010 The most potent stimulus to respiration is:
A. carbon dioxide
B. oxygen
C. potassium
D. sodium
E. nitrogen

A. is correct
85% of the resting respiratory drive results from $CO_2$ stimulation of central chemoreceptors.

3.011 Just as there is some non-uniformity of gas distribution in healthy persons, so there is also some unevenness of capillary blood flow. The position in which unevenness of pulmonary blood flow per unit of lung is greatest is the:
A. supine
B. prone
C. left lateral decubitus
D. trendelenbury
E. vertical

E. is correct
Uneven distribution of blood flow in vertical lung is explained by the hydrostatic pressure differences within the blood vessels. For example: the pressure difference in a vertical 30 cm lung will be 30 cm $H_2O$.

3.012 A patient has arterial blood gases with a $PCO_2$ greater than 70 mmHg. He has narcosis, which means the $CO_2$ center in the medulla no longer controls or drives respiration. In this situation, the patient is breathing because of:
A. hypoventilation
B. acidosis
C. hypoxemia
D. cerebral ischemia
E. alkalosis

C. is correct
Although $CO_2$ is the main stimulus for breathing; this control is lost in this case. Therefore, ventilation decreases (hypoventilation) causing $EPCO_2$ and $FO_2$. Thus peripheral chemoreceptors are stimulated due to hypoxemia also by $EPCO_2$.

3.013 The normal oxygen-carrying capacity of one gram of hemoglobin-A has been found to be:
- A. 0.134 ml
- B. 1.34 ml
- C. 13.4 ml
- D. 134 ml
- E. 1.34 liters

B. is correct
The actual number of ml of $O_2$ that are carried in each 100 ml of blood in chemical combination with hemoglobin (Hb) depends upon the Hb concentration. Each gram of Hb can combine with 1.34 ml $O_2$. If Hb is 15 gm/100 ml then the maximal amount of $O_2$ carried in combination with Hb is 1.34 x 15 = 20 ml $O_2$/100 ml blood.

3.014 In a normal healthy individual, which of the following gases is diffusion limited?
- A. $O_2$
- B. CO
- C. Helium
- D. $CO_2$
- E. $NO_2$

B. is correct
Carbon monoxide is taken up so rapidly by the hemoglobin in the red blood cells that the partial pressure of CO in the capillaries is very low. Thus, the amount of CO entering the body is not limited by the amount of pulmonary blood flow but by diffusion.

**Use the following table to answer question 15.**

| Location | Particle size trapped |
|---|---|
| Nasal Passages | 4-6 micrometers |
| Small bronchioles | 1 - 5 micrometers |
| Alveoli | 7 0. 5 micrometers |

3.015 A buffered solution is able to minimize the deviation of pH resulting from the addition to that solution of a strong acid or a strong base. At the pH of serum (7. 40), the ratio of concentration of sodium bicarbonate to carbonic acid plus dissolved $CO_2$ is:
- A. 1:1
- B. 5:1
- C. 10:1
- D. 20:1
- E. 30:1

D. is correct
From Henderson Hasselbach eg/: pH = pk + log

$$\frac{[HCO_3^-]}{[(0. 03 \text{ m M/l})/ \text{ mmHg}] PCO_2}$$

at pH = 7. 40; $\dfrac{[HCO_3^-]}{PCO_2} = \dfrac{24}{1.2} = \dfrac{20}{1}$

3.016 In deep sea diving:
- A. breathing air at a depth of 100 ft., one may become euphoric because of $O_2$ toxicity.
- B. when breathing air at 100 ft. (4 atms.) the expected $P_1O_2$ would exceed 700 mmHg and prolonged use might be fatal.
- C. when interrupted by a rapid ascent from a depth of 150 ft., may produce the "bends" primarily because of gas bubble formation by $O_2$ and $CO_2$.
- D. helium is used to replace $N_2$ because it is less dense at depths (100 ft. or more) and has less narcotic effect
- E. during descent $N_2$ is slowly removed from the tissues.

D. is correct
Helium is about one half as soluble as $N_2$ so less is dissolved in tissues. In addition, it has a much lower molecular weight and therefore diffuses more rapidly through tissue.

3.017  If one examines partial pressure of gases and lung function in the normal individual:
  A. At sea level, breath air, $P_AO_2$ is approximately 100 mmHg while $P_aO_2$ is about 95 or 97 mmHg
  B. Partial pressure of expired $O_2$ is less than the partial pressure of alveolar $O_2$
  C. Dead space gas is usually a larger volume than expired gas
  D. The sum of partial pressures of arterial gases is approximately equal to the sum of partial pressures of mixed venous gases
  E. A and D

A. is correct
By the time $O_2$ has reached the alveoli, the $PO_2$ is about 100 mmHg. The normal arterial $PO_2$ is around 95-97 mmHg.

3.018  In the work of breathing:
  A. Overcoming compliance of the lung and chest is normally the largest component of total work.
  B. The normal value of airway resistance is 1. 5 l/cm of $H_2O$/sec
  C. Surfactants usually decrease compliance and thus lower the total work required
  D. Energy expenditure is usually more than 5% of total metabolic energy in the quiet resting state
  E. A and C

A. is correct
The lower the compliance the greater the elastic forces that must be overcome during inspiration.

$$\text{Compl.} = \frac{\Delta V}{\Delta P}$$

3.019  In hypoxia(s) or hypoxemia(s):
  A. Arterial-venous $O_2$ difference is greater than normal in stagnant hypoxemia
  B. Partial pressure difference for $O_2$ (arterial-venous) in anemic state is greater than in the normal condition
  C. An arterial hypoxemic patient may display cyanosis
  D. Positive end-expiratory pressure (PEEP) breathing tends to reduce arterial hypoxemia
  E. all of the above

E. is correct
A/B) - In both conditions, increased $O_2$ will be unloaded in the periphery. Therefore, the a-v difference will be greater. Cyanosis is due to the presence of reduced Hb. Breathing against a positive end-expiratory pressure (PEEP) is of value because it recruits previously unventilated alveoli.

3.020  The Ventilation-Perfusion Ratio $V_A$/Q:
  A. The ratio tends to vary several fold in the upright individual
  B. Ratios will be exaggerated in the supine position compared to all erect positions
  C. Ratios for the lung in its entirety exhibit greater variation in the numerator than the denominator
  D. Values are of little use in understanding alveolar arterial gradients for $O_2$
  E. A and C

A. is correct
Because the right ventricle develops a low pressure, blood flow is compromised to the apex of the lung. Most of the blood ejected by the right heart goes to the base of the lung. Thus, in the upright position, more blood than air goes to the base of the lungs and alveoli at the base are relatively hypoventilated.

**Use the following table to answer question 21.**

| Upright | $V_A$ | Q | $V_A$/Q |
|---------|-------|------|---------|
| Apex | .24 | .07 | 3.3 |
| Base | .82 | 1.29 | .63 |

units = (l/min)

3.021 Patients with $V_A/Q$ inequalities often display hypoxemia but have normal or sub-normal $P_aCO_2$ because:
  A. in the physiological range, the $O_2$ and $CO_2$ dissociation curves have different shapes
  B. $CO_2$ diffuses more rapidly than $O_2$
  C. initially the $CO_2$ diffusion gradient is less than the $O_2$ gradient
  D. $CO_2$ has a greater solubility than $O_2$ at body temperature and pH
  E. all of the above

A. is correct
There are differences in the ventilation/perfusion ratio in various parts of the normal lung due to gravity and local changes in the ventilation/perfusion ratio.

3.022 Pulmonary blood:
  A. flow produces essentially no lymph and is so well-controlled and "fine-tuned" that under normal conditions lymph production is negligible or near zero on a daily basis
  B. volume remains a constant value throughout a normal 24 hour period
  C. pressure elevation tends to cause increased pulmonary resistance
  D. flow is influenced by recruitment of blood vessels as well as distension of vessels
  E. C and D

D. is correct
Two mechanisms are responsible for a fall in vascular resistance as pulmonary capillary pressure rises:
1) recruitment - some capillaries are not perfused; when pressure rises these vessels open thereby lowering pressure; 2) distension - at high pressures individual capillaries widen thereby lowering pressure

3.023 At the end of a normal expiration at sea level, the intrapulmonic pressure is greater than:
  A. atmospheric pressure
  B. interpleural pressure
  C. intrapulmonic pressure at the end of inspiration
  D. that at 12,000 ft.
  E. B and D

E. is correct
B) At the end of exp. intrapleural pressure is -2mmHg and intrapulmonic pressure is 0 mmHg.
D) Atmospheric pressure decreases at high elevations and intrapulmonic pressure decreases in a corresponding manner.

3.024 In an otherwise normal man, whose hemoglobin content is 10 gm/100 ml of blood, which of the following is the correct statement:
  A. peripheral chemoreceptors are stimulated by his hypoxemia when he breathes room air
  B. the percentage saturation of hemoglobin in his venous blood is normal while breathing room air
  C. the $O_2$ content of his blood is normal if he breathes 100% $O_2$
  D. the percentage saturation of hemoglobin in his arterial blood is normal when he breathes room air
  E. none of the above

D. is correct
Normal Hb content is 15 g/100 ml. His arterial Hb will be near full saturation as in a normal state; however, his venous Hb will not be due to increased unloading of $O_2$ from the Hb at the periphery causing the decreased Hb content.

3.025  In exercise:
   A.  the Hb-$O_2$ dissociation curve in exercising skeletal muscle is shifted to the right and the arterial venous $O_2$ content difference is diminished.
   B.  the $O_2$ dissociation curve in exercising skeletal muscle (after 15 minutes) is shifted to the left because of the increased production of $CO_2$.
   C.  the initial abrupt increase in respiratory minute volume is the result of an increase in 2,3 DPG
   D.  the initial abrupt increase in respiratory minute volume is because of the increase in $D_L$ for $O_2$ and $CO_2$ together with an increase in cardiac output
   E.  none of the above

E. is correct
A) The arterial venous $O_2$ difference is increased. B) Curve is shifted to the right. C) DPG does not play a role in shifting the $O_2$ dissociation curve in exercise. D) Respiratory minute volume is the amount of new air moved into the respiratory passage each minute. $D_L$ does not affect respiratory minute volume; especially at the first abrupt increase of respiratory minute volume.

3.026  The diffusion capacity of the lung ($D_L$) is:
   A.  greater for $O_2$ than for $CO_2$
   B.  affected by surface area available for diffusion and alveolar recapillary membrane thickness
   C.  decreased during exercise
   D.  not influenced by the alveolar to pulmonary capillary partial pressure gradient of a particular gas
   E.  B and C

B. is correct
$D_L$ is affected by area, thickness, diffusion properties of the tissue sheet and the gas concerned.

$$D_L = \frac{V_{gas}}{P_1 - P_2}$$

3.027  $CO_2$ transport:
   A.  the mixed venous to alveolar gradient for $CO_2$ is small compared to the gradient one normally sees for $O_2$ across the lung
   B.  may be impeded by breathing 100% $O_2$
   C.  would tend to be retarded during histotoxic hypoxemia
   D.  the primary net transport material is $HCO_3^-$ carried in the plasma
   E.  all of the above

E. is correct
$P_aO_2 = 40$ $P_AO_2 = 100$; High $PO_2$ can impede $CO_2$ transport through haldane effect; $P_aCO_2 = 45$ $P_ACO_2 = 40$; Histotoxic hypoxemia impairs ability to transport $O_2$ and $CO_2$.

3.028  At altitude:
   A.  a subject initially (first hour) displays both hypoxemia and hypocapnia
   B.  a long-term resident presents an increased $O_2$ capacity
   C.  a new arrival exhibits an increase in $P_{50}$
   D.  a person has decreased respiratory minute volume
   E.  A and B

E. is correct
At altitude - a subject increases his alveolar $PO_2$ and blows off $CO_2$ by hyperventilating. Long-term residents increase the amount of oxygen than can be combined with hemoglobin.

3.029.  Carbon monoxide:
   A.  combines with Hb much more avidly than $O_2$
   B.  Shifts the $O_2$ dissociation curve to the right while it lowers the $O_2$ capacity of the hemoglobin
   C.  is an odorless and tasteless gas which exerts very low partial pressure in arterial blood, and in addition, fails to stimulate either f or $V_T$
   D.  diffuses across the alveolar-pulmonary capillary membrane more slowly than $O_2$
   E.  all of the above

E. is correct
Oxygen has a diffusion capacity 1. 23 times that of carbon monoxide.

3.030 When the $V_A/Q$ of a lung unit increases:
  A. [H$^+$] of blood (perfusing) this unit increases
  B. P$_A$O$_2$ increases
  C. P$_A$H$_2$O increases
  D. P$_A$CO$_2$ increases
  E. none of the above

B. is correct
When the ventilation perfusion ratio increases alveolar O$_2$ concentration increases.

3.031 Surfactant:
  A. is probably not a problem in adult respiratory distress syndrome
  B. tends to stabilize alveoli
  C. is produced by Type I alveolar cells
  D. tends to keep alveoli moist and free from drying out
  E. may cause pulmonary edema

B. is correct
Surfactant tends to stabilize alveoli and helps keep alveoli dry. It is produced by type II alveolar cells.

3.032 Hypoxic vasoconstriction of pulmonary vessels is due to:
  A. a reflex arc through C3-C5 of the spinal cord
  B. activation of the Hering-Breuer reflex
  C. reduced PO$_2$ in alveolar gas
  D. PO$_2$ of mixed venous blood as it transits the lung
  E. increased PO$_2$ in alveolar gas

C. is correct
Pulmonary vesicles vasoconstrict during hypoxia thereby reducing perfusion to the alveoli with low PO$_2$.

3.033 With regard to chemoreceptors:
  A. they are located in the area of the 4th ventricle perhaps bathed by cerebrospinal fluid (CSF) and peripherally on the aorta and carotid arteries.
  B. central chemoreceptors respond to PO$_2$ of blood or CSF
  C. peripheral receptors monitor P$_a$O$_2$, P$_a$CO$_2$, [H$^+$] and temperature
  D. an increase [H$^+$] in arterial blood causes both sets of receptors to augment respiratory effort
  E. none of the above

E. is correct
Peripheral chemoreceptors are not bathed in CSF, don't monitor temperature control receptors don't respond to changes in H$^+$, but do respond to changes in pH (CO$_2$).

3.034 Normally there is an alveolar-to-arterial PO$_2$ difference of about 5-7 mmHg because:
  A. the diffusion capacity of O$_2$ is less than that for CO$_2$.
  B. the ventilation/perfusion ratio is the same everywhere in the lungs.
  C. no gas exchange occurs in the large airways.
  D. there is a small right-to-left shunt.
  E. the solubility of O$_2$ is greater in air than in blood

D. is correct
The pressure of A-a gradient for oxygen indicates the presence of a shunt.

3.035 With advanced age:
  A. total lung capacity increases
  B. critical closing volume of the lung normally increases
  C. functional residual capacity remains approximately 40% of total lung capacity
  D. in the lung, O$_2$ saturation of hemoglobin tends to be increased
  E. B and D

B. is correct
In young normal subjects, the critical closing volume is 10% of vital capacity, with age it increases to about 40% of vital capacity.

3.036 Interpleural pressure:
A. is "more negative" (further below atmospheric) at diaphragmatic sites than apical areas of the lung, in the erect individual
B. gradient is abolished in astronauts during circular earth orbit
C. always remains less than atmospheric pressure throughout life
D. normally aids venous return to the right heart during expiration
E. none of the above

B. is correct
The interpleural pressure gradient is abolished in astronauts during circular earth orbit.

3.037 Dead space:
A. is normally about . 15l/minute and increases during exercise
B. gas is the first to reach the functional residual capacity
C. tends to elevated by high $V_A$/Q ratio
D. has a $PO_2$ similar to inspired air and a $CO_2$ partial pressure nearly identical to expired gas
E. none of the above

C. is correct
Patients with $V_A$/Q inequalities ventilate in excess. This is necessary because lung units with high $V_A$/Q ratios are insufficient at eliminating $CO_2$. These units make up alveolar dead space.

3.038 The loss of pulmonary surfactant is likely to cause all of the following EXCEPT:
A. atelectasis
B. an increase in the work of expanding the lung.
C. transudation of fluid from capillaries into alveoli.
D. an intrapleural pressure that is more negative than normal during inspiration.
E. an increase in alveolar ventilation

E. is correct
The loss of surfactant results in areas of collapsed alveoli, atelectasis, low compliance, and alveoli filled with transudate. These changes result in a decrease in alveolar ventilation not an increase.

3.039 As pulmonary arterial blood traverses the lungs, the following happen(s):
A. hematocrit decreases
B. $P_{50}$ gets smaller
C. $O_2$ quickly combines with Hb
D. Cl⁻ moves out of the RBC
E. all of the above

E. is correct
As pulmonary arterial blood flows across the lung, the lungs filter the blood and trap cells to decrease hematocrit, oxygen combines with Hb, chloride moves out and bicarbonate moves in.

3.040 In a normal subject breathing air sea level, the partial pressure of water vapor ($PH_2O$) in alveolar gas:
A. is directly related to alveolar ventilation ($V_A$) and thus would be elevated at high altitude
B. is less than the alveolar partial pressure of $CO_2$
C. is the same as $PH_2O$ of arterial blood
D. would decrease if body temperature increases
E. none of the above

C. is correct
The partial pressure of water vapor ($PH_2O$) is the same as $PH_2O$ of arterial blood.

3.041 In addition to gas exchange, a lung:
A. acts as a filter and may trap blood clots
B. in an adult, has lost the ability to synthesize protein such as collagen and elastin
C. removes or inactivates some blood-borne hormones as they transit the pulmonary circuit
D. converts angiotensin-I to angiotensin-II as well as producing erythropoietin
E. A and C

E. is correct
As mentioned earlier the lung filters the blood and can trap blood clots and can modify blood borne substances such as bradykinin, prostaglandins, etc.

3.042. Which of the following statements regarding oxygen transport is (are) true?
  A. if an individual's blood contains fetal hemoglobin, this situation leads to reduced unloading of $O_2$ at peripheral tissues at a standard partial pressure of $O_2$
  B. in a patient with an elevated $P_aCO_2$ (46 mmHg) the arterial $O_2$ content will be less than when his $P_aCO_2$ is 40 mmHg
  C. physically dissolved $O_2$ is similar in a patient breathing air or in a patient with half of his hemoglobin combined with carbon monoxide
  D. an individual during work exhibits an increase cardiac output and an increased arterial venous $O_2$ content difference, thus supplying the greater tissue needed for oxygen
  E. all of the above

E. is correct
All the statements are correct.

3.043 A normal person at sea level doubles alveolar ventilation, the $O_2$ content of the blood perfusing his lungs will:
  A. double
  B. increase 1. 5 times
  C. remain the same
  D. decrease to one half
  E. increase 3 fold

C. is correct
$O_2$ content will not change because Hb is saturated (the $O_2$ dissociation curve is at the plateau).

3.044. In a normal, erect individual, the apical region of the lungs compared to the base of the lung show:

| Ventilation/Unit Vol. | Blood Flow/Unit Vol. | $V_A$/Q |
|---|---|---|
| A.  less | less | greater |
| B.  less | greater | lower |
| C.  less | less | lower |
| D.  greater | greater | greater |
| E.  greater | less | greater |

A. is correct
Both ventilation and perfusion are decreased at the apex compared to the base of the lungs. The ventilation perfusion ratio is greater at the apex than at the base of the lungs.

3.045 Partially compensated respiratory acidosis is associated with an increase (from normal) of all the following EXCEPT:
  A. hydrogen ion concentration
  B. buffer base
  C. standard bicarbonate
  D. partial pressure of $CO_2$
  E. ratio of bicarbonate to carbonic acid concentration

E. is correct
The ratio of bicarbonate to carbonic acid concentration decreases during partially compensated respiratory acidosis.

3.046 After maximal expiration, relaxation pressure will be due to:
  A. the additive recoil of the lungs and chest wall
  B. only the lungs
  C. only the chest wall
  D. the chest wall minus that of the lungs
  E. the chest wall plus that of the lungs

D. is correct
To correctly answer this question requires an understanding of the relaxation pressure volume diagram.

3.047  The alveolar-arterial gradient for $O_2$ in man is:
A. normally about 15 mmHg
B. enlarged in a left to right shunt
C. in part, due to $V_AQ$ imperfections
D. essentially identical to the gradient seen for $CO_2$

C. is correct
The concentration of $O_2$ in the alveolar is regulated by the rate of blood flow and the rate of ventilation. Thus, the concentration of $O_2$ and thus the A - a gradient is determinate by $V_A/Q$ ratios.

3.048  All of the following are correct for pulmonary airway resistance, EXCEPT:
A. is normally responsible for about 30% of the metabolic expenditure involved in respiration
B. is less at high lung volumes compared to low lung volumes
C. is greatest in the intermediate sized bronchi (5-8 generation)
D. is elevated in an individual undergoing an acute asthmatic episode or attack
E. is increased by surfactant

E. is correct
Surfactant, a lipid, decreases pulmonary airway resistance. If the surface tension is not kept low during expiration, the alveoli will collapse due to the Law of Laplace. Surfactant also prevents pulmonary edema by preventing transudation of fluid from the blood to alveoli.

3.049  Lung compliance:
A. is decreased by the presence of surfactant
B. is decreased by atelectasis
C. is increased in the fibrotic patient
D. for the lung and chest together is approximately 1. 0 l/cm $H_2O$
E. is decreased in emphysema

B. is correct

$$compliance = \frac{WV}{WP}$$

Atelectasis is the collapse of the alveoli; therefore, with a given WV the WP will be less, and thus, compliance will be less.

3.050  The diffusion capacity of the lung is:
A. approximately 250 ml/min/mmHg for $O_2$
B. remarkably uniform throughout its structure
C. usually measured in a clinical lab utilizing a nitrogen meter
D. increased during exercise, in part, due to increased pulmonary capillary blood volume as well as increased $V_A$
E. is unchanged during exercise

D. is correct

$$D_L = \frac{V_{gas}}{P_1 - P_2}$$

$P_1$ and $P_2$ equal the partial pressure of alveolar and capillary blood, respectively. If $V_A$ and $P_2$ increases then $D_L$ increases.

3.051  In peripheral systemic capillaries more $O_2$ can be given up by hemoglobin at a given $PO_2$ if there is an increase in the:
A. $PCO_2$ of the blood
B. hematocrit
C. temperature of the blood
D. pH of the blood

A. is correct
Due to the Haldane effect, greater $PCO_2$ facilitates the unloading of $PO_2$.

3.052  In the normal individual, the medullary respiratory chemoreceptors are stimulated by all of the following, EXCEPT:
A. fall in $P_aO_2$
B. rise in $P_aO_2$
C. decreased arterial [$H^+$] concentration
D. decrease in cerebral spinal fluid pH

C. is correct
$H^+$ cannot cross the blood-brain barrier

3.053 Carbon dioxide:
A. net transport is principally via carbamino hemoglobin
B. partial pressure elevation decreases brain perfusion
C. is associated with euphoria in underwater divers
D. has a lower partial pressure in high altitude natives
E. in toxic inhalations causes hypoventilation

D. is correct
If the atmospheric pressure decreases, the $PCO_2$ decreases in a corresponding manner.
$PCO_2 = P_{ATM} \cdot FCO_2$.

3.054 All of the following are reduced in emphysema EXCEPT:
A. forced vital capacity
B. timed vital capacity
C. % of forced vital capacity exhaled in one second
D. forced mid-expiratory flow
E. total lung capacity

E. is correct
Although obstructive disorders decrease air flow indicators, total lung capacity is not changed.

3.055 In which of the following patients would we least likely observe cyanosis?
A. arterial hypoxemia
B. stagnant hypoxemia
C. $V_A/Q$ imbalances
D. anemia
E. gaseous diffusion difficulties (lung)

D. is correct
Cyanosis is due to an increase in reduced Hb (non-oxygenated Hb). Anemia is a reduction in the concentration of Hb. Therefore, the Hb present will be fully saturated with $O_2$ and not be reduced.

3.056 A man is rescued from a burning building just as he loses consciousness. After breathing pure outdoor air for five (5) minutes, which of the following would be observed?
A. respiration is not stimulated, although he is still hypoxic
B. still stimulated by hypoxia
C. severely depressed by hypoxia of the aortic and carotid bodies
D. all his hemoglobin has become full oxygenated
E. his hematocrit has elevated 80%

A. is correct
The person is probably suffering from carbon monoxide poisoning with the CO binding to most of his hemoglobin and thereby inducing hypoxia but not stimulating respiration. The patient needs to be treated with pure oxygen so that the CO will be displaced since breathing oxygen at the low pressure of the atmosphere will take longer to dispence the CO from the Hb. The patient will also benefit from simultaneous administration of $CO_2$ because $CO_2$ stimulates the respiratory center.

3.057 A subject is on a mountain at 18,000 ft. ($P_B$ = 380 mmHg), $H_2O$ vapor partial pressure is 47 mmHg at 37°C and he inspires normal air. Partial pressure of $O_2$ saturated inspired (tracheal) air will be:
A. 47 mmHg
B. 70 mmHg
C. 80 mmHg
D. 95 mmHg
E. 333 mmHg

B. is correct
$PO_2 = (O_B - 47) FO_2$; $PO_2 = (380 - 47) 0.21$;
$PO_2 = 70$ mmHg

3.058 All of the following are respiratory control centers in the pons and medulla, EXCEPT:
A. pontine respiratory group
B. dorsal medullary respiratory group
C. ventral medullary respiratory group
D. a central pattern generatory
E. a dorsolateral medullary respiratory group

E. is correct
A dorsolateral medullary respiratory group does not exist.

3.059 All of the following are peripheral sites of sensory information to the respiratory centers, EXCEPT:
- A. joints
- B. muscles
- C. long bones
- D. lung reception
- E. carotid bodies

C. is correct
Although joints act as peripheral inputs of sensory information to the respiratory centers long bones do not.

3.060 All of the following are functions of surfactant, EXCEPT:
- A. prevents alveoli closing during expiration
- B. contributes to the immune response against infection
- C. increases pleural pressure
- D. maintains alveolar stability during exhalation
- E. acts to waterproof alveoli

C. is correct
Surfactant does not increase pleural pressure

3.061 Lung function is altered in morbid obesity by all the following, EXCEPT:
- A. increasing the work of breathing
- B. increasing chest wall compliance
- C. increased inspiratory reserve volume
- D. decreased expiratory reserve volume
- E. decreased functional residual capacity

C. is correct
Inspiratory reserve volume will not be changed because inspiration is active and respiratory muscles are involved in labored breathing.

3.062 The percentage of hemoglobin saturated with oxygen will increase, EXCEPT:
- A. the arterial $PCO_2$ is increased
- B. the hemoglobin concentration is increased
- C. the temperature is increased
- D. the arterial $PO_2$ is increased
- E. the arterial pH is increased

B. is correct
Since hemoglobin is normally around 97% saturated only more hemoglobin will increase the percentage of oxygen saturated hemoglobin.

3.063 The reaction of carbon dioxide with water to form carbonic acid occurs:
- A. mainly in the interstitial fluid
- B. mainly in blood plasma
- C. mainly in red blood cells (RBCs)
- D. equally well in interstitial fluid, plasma and RBCs
- E. equally well in plasma and RBCs

C. is correct
Because carbonic anhydrase is only present in the RBC.

3.064 Hydrogen ions dissociated from carbonic acid in venous blood:
- A. are quickly buffered by hemoglobin molecules
- B. are mainly buffered by bicarbonate ions in plasma
- C. largely remain free and cause a significant pH shift to the acid side
- D. largely diffuse out of red cells and are buffered by albumin and other plasma protein molecules
- E. quickly combine with hydroxyl molecules to form water

A. is correct
$H^+ + HbO_2 = H \times Hb + O_2$

3.065 The largest amount of $CO_2$ is transported by the blood as:
- A. $CO_2$ in plasma
- B. Bicarbonate in the red blood cells
- C. $H_2CO_3$ in plasma
- D. carbamino-$CO_2$
- E. bicarbonate ion in the plasma

E. is correct
Most of the $CO_2$ produced by metabolism is carried by the plasma in the form of bicarbonate. The remainder is carried as carbamino compounds and dissolved $CO_2$ in the blood

3.066 When blood carbon dioxide content increases and oxygen content decreases in peripheral tissues:
   A. red blood cells increase only the fraction of carbon dioxide carried as bicarbonate in plasma but not the fraction carried as carbamino compounds
   B. the affinity of hemoglobin for oxygen decreases
   C. the hemoglobin molecule develops a greater tendency to dissociate hydrogen ions
   D. the amount of chloride within the red cells decreases
   E. the oxygen dissociation curve is shifted to the left

B. is correct
This is due to the Haldane effect. The binding of oxygen with hemoglobin tends to displace carbon dioxide from the blood.

3.067 The following statements on decompression sickness are true, EXCEPT:
   A. the amount of gas dissolved in solution is directly proportional to the ambient pressure and the amount of time spent at the given depth
   B. the primary gas of concern is nitrogen, primarily because of its high fraction in the inspired gas and of its free diffusibility in body tissues
   C. the effects are those of nitrogen bubbles coming out of solution
   D. treatment consists of the administration of high partial pressures of oxygen to displace the dissolved nitrogen gas
   E. the use of inspired ambient air pumped from the surface allows divers to descend to moderate depths without fear of decompression sickness

D. is correct
Treatment of decompression sickness is by recompression which reduces the volume of bubbles and forces them back into solution.

3.068 An increase in respiratory dead space will cause a:
   A. rise in arterial oxygen partial pressure
   B. fall in alveolar carbon dioxide partial pressure
   C. fall in alveolar oxygen partial pressure
   D. fall in hydrogen ion concentration in arterial blood
   E. fall in arterial carbon dioxide partial pressure

C. is correct
Since dead space is increased by a higher ventilation perfusion ratio that will result in alveolar oxygen partial pressure.

3.069 The diagram shows the pressure volume loop of the same lung. One loop was obtained by inflating the lung with air. The other was obtained by filling the lung with saline. Compared with air filled lung, the saline filled lung shows:
   A. complete emptying at zero inflation pressure
   B. a higher compliance
   C. a larger volume at maximal inflation
   D. greater hysteresis
   E. less resistance to air flow

B. is correct
The saline filled lungs have a much higher compliance and less hysteresis than the air-filled lung.

3.070 The oxygen dissociation curve for hemoglobin is shifted to the right by:
   A. decreased $O_2$ tension
   B. decreases diphosphoglycerate
   C. increased pH
   D. increased $CO_2$ tension
   E. increased $N_2$ tension

D. is correct
An increase in $PaCO_2$ shifts the $HbO_2$ dissociation curve to the right indicating a decrease in the affinity of $O_2$ for Hb. Thus, the A-V $O_2$ difference increases.

3.071 All of the following statements concerning a patient suffering from CO poisoning are true, EXCEPT:

A. the oxygen carrying capacity of the blood is decreased

B. $PO_2$ of the arterial blood is lower than normal

C. the dissociation curve of oxyhemoglobin shifts to the left

D. the patient experiences minimal symptoms when the blood levels of carboxyhemoglobin are less than 30%

E. the patient can be effectively treated with 100% oxygen although such treatment may be inadvisable

B. is correct

Hb avidly binds CO over $O_2$. However, $PO_2$ is affected by dissolved $O_2$ and not the amount bound to Hb.

3.072 Bilateral section of the vagi in an anesthetized dog:

A. stops diaphragmatic movements while allowing intercostal activity to continue

B. speeds the breathing by interrupting vagal inhibitory fibers

C. speeds the breathing if it has been previously depressed by severe hypoxia

D. slows the frequency of breathing although tidal volumes may increase

E. affects breathing movements only if the peripheral chemoreceptors were previously being stimulated

D. is correct

The Hering-Breuer reflex causes increased firing of lung stretch receptor. Impulses are sent via the vagus to the apeustic center where inspiration is inhibited. If the vagi are cut inspiration is not inhibited reading to a longer inspiratory period and a great TV and a delayed expiration.

3.073. The figure below is the spirometric tracing of a healthy individual breathing normally and executing a vital capacity maneuver. The function residual capacity is represented by:

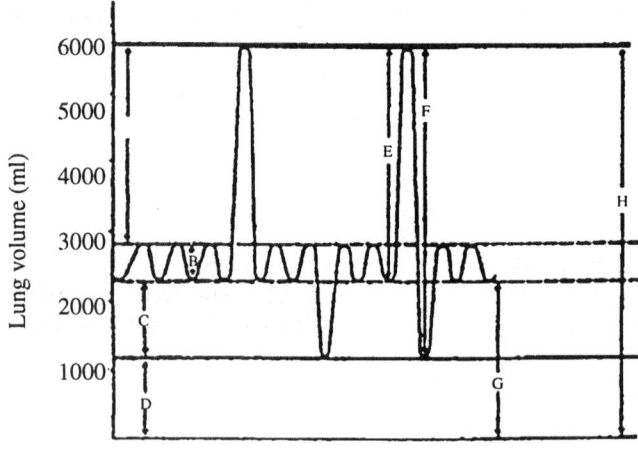

A. A

B. B plus C plus D

C. A plus E

D. B plus F

E. C plus G

E. is correct

G represents FRC. FRC = RV + ERV

3.074 You are monitoring the arterial $PCO_2$ and pH of an anesthetized patient undergoing surgery. Ventilation is controlled by mechanical respirator. The initial values are normal ($PCO_2 = 40$ mmHg, pH = 7. 42). When ventilation is doubled, which of the following occurs?

|  | $PCO_2$ | pH |
|---|---|---|
| A. | decreased | decreased |
| B. | increased | decreased |
| C. | decreased | increased |
| D. | increased | increased |
| E. | decreased | no change |

C. is correct
If ventilation increases $CO_2$ is blown off and therefore, pH decreases.

3.075 During exercise:
A. $PO_2$ in venous blood would increase
B. $PCO_2$ in venous blood would increase
C. ventilation/perfusion ratio would decrease
D. overall peripheral resistance increases
E. the ventilation/perfusion ratio at the base of the lungs is higher than the apices

B. is correct
Due to increased metabolic activity of muscle cells, increased concentration of $CO_2$ would diffuse into the venous blood.

3.076 Carbon dioxide:
A. increases oxygen binding by hemoglobin
B. is transported in the arterial blood principally as molecular $CO_2$ in physical solution
C. when added to the blood, causes almost no change in the concentration of bicarbonate ion
D. enters capillary blood by passive diffusion from the tissues
E. has a very small partial pressure in capillary blood

D. is correct
Diffusion of $CO_2$ through tissue is about 20 times faster than that of $O_2$.

3.077 Which of the following listed below most clearly approximates the respiratory quotient (RQ) of an animal oxidizing fat?
A. 1. 0
B. 0. 9
C. 0. 8
D. 0. 7
E. 0. 6

D. is correct
Fats - R = 0. 7; Protein - R = 0. 8; Carb. - R = 1. 00

3.078 If tidal volume is maintained constant, physiological dead space is increased when:
A. $P_AO_2$ and $P_ACO_2$ increase equally
B. the sum of $P_AO_2$ and $P_ACO_2$, equals $P_IO_2$
C. $P_ACO_2$ and $P_ECO_2$ increase equally
D. the $P_ACO_2$ and $P_ECO_2$ difference decreases
E. the $P_ACO_2$ and $P_ECO_2$ difference increases

E. is correct
Physiologic dead space increases in the pressure of ventilation/perfusion inequality.

3.079 The metabolic production of carbon dioxide is halved (as in hypothermia). Which of the following must also be halved if arterial carbon dioxide partial pressure is to remain in the same:
A. tidal volume
B. total ventilation
C. alveolar ventilation
D. carbon dioxide output
E. physiological dead space

C. is correct

$$V_A = \frac{VCO_2}{P_aCO_2}$$

Alveolar ventilation is the amount of air reaching the alveoli per minute. If $CO_2$ production is reduced by half, alveolar ventilation must also be halved to keep the $PaCO_2$ the same.

3.080  Following infusion of lactic acid into the blood of a normal subject which results in a fall in arterial pH to 7.35, you would expect all of the following EXCEPT:
- A.  an increase in ventilation
- B.  a fall in the pH of cerebrospinal fluid
- C.  a rise in arterial pH
- D.  a fall in arterial $PCO_2$
- E.  an increase in ventilation/perfusion ratio

C. is correct
Arterial pH decreases following the infusion of lactic acid.

3.081  Assume the following values: partial pressure of $CO_2$ in alveolar air = 40 mmHg, partial pressure of $CO_2$ in expired air = 30 mmHg, tidal volume = 800 ml. The physiological dead space would be:
- A.  150 ml
- B.  200 ml
- C.  250 ml
- D.  400 ml
- E.  450 ml

B. is correct
Bohr equation should be used.

$$P_E CO_2 \times V_T = P_A CO_2 \times (V_T - V_D)$$

$$\text{or } V_T - V_D = \frac{P_E \times V_T}{P_A CO_2}$$

$$\therefore \frac{30 \times 800}{40} = 800 - V_D$$

$$VD = 800 - 600$$
$$= 200$$

3.082  In the lungs, fluid normally flows from the tissue spaces into the capillaries throughout their length because:
- A.  osmotic pressure is low
- B.  capillary blood pressure is low
- C.  blood volume is low
- D.  tissue pressure is low
- E.  the alveolar epithelial membrane tends to pull away from the capillary membrane

B. is correct
The lungs have a very low capillary pressure of around 7 mmHg compared to the higher functional capillary pressure of around 17 mmHg elsewhere in the body.

3.083  The position in which unevenness of pulmonary blood flow per unit of lung is greatest is the:
- A.  supine
- B.  prone
- C.  left lateral decubitus
- D.  trendelenburg
- E.  vertical

E. is correct
When the thorax is in the upright position, more air and blood go to the base of the lungs than to the apex. More blood than air, however, goes to the base. Thus, alveoli at the base are relatively hypoventilated for the amount of blood flow. Alveoli at the apex receive too much air for the amount of blood perfusing them.

3.084  In a normal 70 Kg man, arterial $PO_2$ will be highest under which of the following conditions?

|  | Alveolar Ventilation | Blood Hemoglobin Content (g/dl) | $PO_2$ of Inspired Air (mmHg) |
|---|---|---|---|
| A. | 7 | 18 | 150 |
| B. | 3 | 7 | 80 |
| C. | 7 | 7 | 80 |
| D. | 3 | 20 | 150 |
| E. | 7 | 7 | 300 |

A. is correct
(1. 39 x Hb x Sat/100) + 0. 003mmHg $PO_2$
(1. 39 x . 8 x 97/100) + 150 (0. 003) mmHg $PO_2$
25. 02 x . 97 +0. 45 = 24. 7 mmHg $O_2$

Only "A" has near "normal" values for alveolar ventilation, hemoglobin content, and inspired air. Choices B and D exhibit hypoventilation. Choices B, C and E exhibit low hemoglobin content. Thus choice A will produce the highest arterial $PO_2$ levels.

3.085 When a man inspires maximally, closes his glottis, and tries to expire as hard as he can (a Valsalva maneuver), all of the following occur EXCEPT:
   A. intrapleural pressure is positive
   B. respiratory muscles and the recoil of the lungs both act in the same direction
   C. gas in the lungs is under positive pressure
   D. venous return of blood to the right side of the heart is impeded
   E. thoracic wall vessels dilate

E. is correct
Increased intrathoracic pressure during a forced expiration with the glottis closed results in compression of thoracic wall vessels leading to decreased venous return to the heart.

3.086 Functions of the lungs that help to protect the brain from agents that could interfere with its function include all of the following EXCEPT:
   A. metabolism of serotonin (5-hydroxytryptamine) released into the blood by platelets at sites of injury
   B. trapping of small emboli in the pulmonary capillaries
   C. serving as a reservoir for the left ventricle when the output from the right ventricle is temporarily diminished
   D. elimination of $CO_2$ from the blood
   E. formation of methemoglobin

E. is correct
Formation of methemoglobin does not protect the brain from toxins.

3.087 Which of the following causes of brain hypoxia would most strongly stimulate the aortic and carotid chemoreceptors?
   A. carbon monoxide poisoning
   B. severe anemia
   C. formation of methemoglobin
   D. a marked decreased in the pulmonary diffusing capacity
   E. acute respiratory alkalosis

D. is correct
The peripheral chemoreceptors respond to: fall in $PO_2$, increased $H^+$, and $CO_2$. A, B, and C all reduce the concentration of Hb that can combine with $O_2$. A reduction in Hb which can combine with $O_2$ does not stimulate the peripheral chemoreceptors. A decreased pulmonary diffusion capacity will decrease $PO_2$ and stimulate the peripheral chemoreceptors.

3.088 During exercise, pulmonary perfusion:
   A. becomes more uniform, apex to base, because of increased cardiac output
   B. decreases at the apex because of an increase in the pulsate nature of the flow
   C. is less well matched to ventilation distribution because, at increased respiratory rates, the vertical gradient in ventilation is decreased
   D. increases proportionately less than pulmonary arterial pressure, since the pulmonary arteries are not capable of significant vasodilatation
   E. becomes slightly but significantly less than systemic perfusion because of increased flow through left-to-right shunts

A. is correct
Remember the pressure difference in a normal non-exercising lung:
Apex: $P_A > P_a > P_v$
Middle: $P_a > P_A > P_v$
Base: $P_a > P_v > P_A$
A portion of the lung will only be perfused if $P_a > P_v > P_A$. During exercise CO increases and Pv increase, therefore the appex and middle are perfused.

**EXTENDED MATCHING** Questions 89-92.

A. Adult respiratory distress
B. Atelectasis
C. Bronchial asthma
D. Bronchiectasis
E. Bronchopneumonia
F. Chronic bronchitis

G. Diffuse interstitial lung disease
H. Emphysema
I. Cystic fibrosis
J. Lung abscess
K. Pulmonary edema
L. Pulmonary embolism

3.089  This disorder is often associated with left heart failure and mitral valve disease.

K. is correct
The pathological accumulation of fluid in the lung is called pulmonary edema and is often associated with left heart failure or mitral valve stenosis.

3.090  This disorder is associated with greatly increased lung compliance and marked dyspnea.

H. is correct
Emphysema is a disease where large numbers of alveoli are destroyed.

3.091  This disease is associated with bronchial smooth muscle spasm that is effectively treated by epinephrine.

C. is correct
Bronchiolar asthma is caused by severe constriction of the bronchioles. Epinephrine counteracts the effects of asthma.

3.092  This disorder is characterized by excessive mucus secretion which is very thick and obstructs airways.

I. is correct
Cystic fibrosis is a genetic defect that causes thick and viscous mucus to be secreted in the pulmonary and GI systems.

**Use the following graphs to answer questions 93-95**

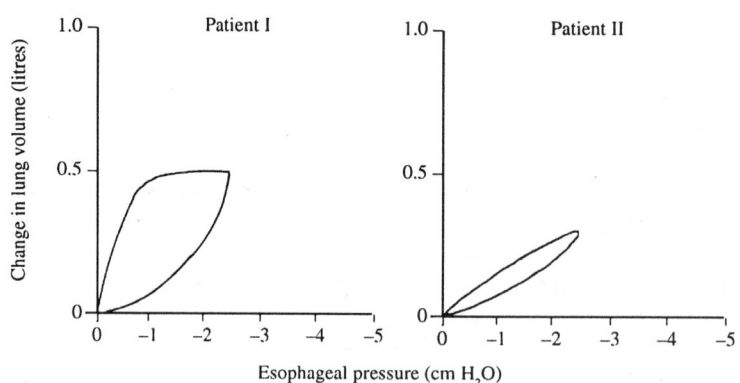

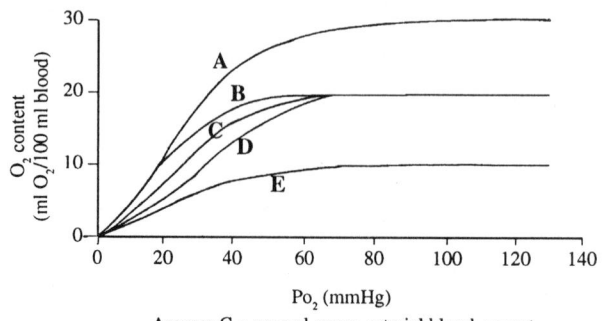

Po$_2$ (mmHg)
Assume C = normal curve, arterial blood, ar rest

3.093 The two graphs above represent tracings of esophageal pressure (with respect to atmospheric pressure) plotted against simultaneous changes in lung volume during quiet breathing at a respiratory rate of 15/min. Which of the following statements is correct?
    A. pulmonary compliance is greater in patient I
    B. pulmonary compliance is greater in patient II
    C. insufficient information is given to estimate pulmonary compliance
    D. tidal volume is greater in patient II
    E. insufficient information is given to estimate tidal volume

A. is correct

3.094 $O_2$ dissociation curve of blood leaving muscle during exercise.

D. is correct

3.095 $O_2$ dissociation curve of blood from a patient with anemia.

E. is correct

**EXTENDED MATCHING** - Questions 96-100

    A. inspiratory capacity
    B. functional residual capacity
    C. vital capacity
    D. inspiratory reserve volume
    E. expiratory reserve volume
    F. residual volume
    G. tidal volume
    H. total lung capacity

3.096 The maximum volume of gas that can be exhaled following normal resting expiration is termed:

E. is correct

3.097 The volume of gas contained in the lungs at the end of maximal inspiration is termed:

H. is correct

3.098 The functional residual capacity minus the expiratory reserve volume is termed:

D. is correct

3.099 The amount of air one can expel from the lungs after a maximal inspiration:

C. is correct

3.100 The amount of air left in the lungs following a normal expiration is termed:

B. is correct

# SECTION 4:   RENAL PHYSIOLOGY

4.001  All of the following statements concerning the renal corpuscles are true EXCEPT:
   A. a basal lamina represents the middle element of the filtration barrier
   B. the endothelium of the capillary loops is fenestrated
   C. the slit membranes prevent the passage of anything equal to or larger than serum albumen
   D. everything except the largest protein molecules can pass through the basal lamina
   E. the foot processes of the podocytes close the fenestrations of the capillaries

E. is correct
The glomerulus is composed of an afferent and efferent arteriole; intervening lined by endothelial cells (tuft); the outer surface of the capillaries which is covered by epithelial cells (podocytes) which are continuous with the epithelium of Bowman's Space and proximal tubule; the mesangium and basement membrane.

4.002  In normal conditions, the portion of the renal tubules which absorbs the largest amount of sodium is:
   A. proximal convoluted tubule
   B. descending portion of Henle's loop
   C. ascending portion of Henle's loop
   D. distal convoluted tubule
   E. collecting tubules

A. is correct
Eighty percent of the glomerular ultrafiltrate is actively absorbed in the proximal tubule.

4.003  The bicarbonate buffer system has a pK of 6.1. If the percent of buffer in the form of $H_2CO_3$ is 50% and the percent of buffer in the form of $HCO_3$ is 50%, the pH of the solution will be:
   A. 7.4
   B. 7.1
   C. 6.4
   D. 6.1
   E. not possible to determine from the data given

D. is correct
According to the Henderson-Hasselbach equation
$$pH = 6.1 + \log \frac{(HCO_3)}{(H_2CO_3)}$$
If both numerator and denominator are equal the pH will remain 6.1.

4.004  The water content of plasma is approximately:
   A. 30%
   B. 45%
   C. 60%
   D. 75%
   E. 90%

E. is correct
The water content of plasma is closest to 90%.

4.005  The rate of glomerular filtration (GFR) can accurately be determined (in man and animals) by measuring the renal clearance of the polysaccharide inulin. The clearance of normal young males averages:
   A. 125 ml/min
   B. 125 mg/min
   C. 125 mg/ml
   D. 125 ml/mg
   E. 125 min/ml

A. is correct
The glomerular filtration rate (GFR) is normally about 125 ml/min and inulin is measured by the clearance of exogenous inulin.

4.006 All the following are correct for inulin EXCEPT:
A. is used to measure GFR
B. does not affect GFR
C. is neither secreted nor absorbed
D. is manufactured in the kidney
E. is not metabolized by the kidney

D. is correct
Inulin is an exogenous substance that is not manufactured or metabolized by the kidney.

**Use the following to answer questions 7 and 8:**

A. renal artery
B. afferent arteriole
C. glomerular capillary
D. efferent arteriole
E. peritubular capillary

4.007 Constriction causes a decrease in renal blood flow and an increase in glomerular filtration rate.

D. is correct
Constriction of the efferent arteriole increases glomerular pressure and GFR and decreases overall renal blood flow.

4.008 Increase in hydrostatic pressure causes decrease in net sodium reabsorption.

E. is correct
The peritubular capillary is a low pressure capillary system and sodium must be actively transported across the basolateral (peritubular) membrane.

**Use the following to answer question 9 and 10:**

A. potassium
B. chloride
C. creatinine
D. urea
E. glucose

4.009 Clearance is about equal to glomerular filtration rate in humans.

C. is correct
Creatinine is an endogenous metabolite formed from phosphorylcreatine. Creatinine is excreted mainly by filtration and its clearance is used to measure GFR.

4.010. Reabsorbed by an active, Tm-limited transport mechanism.

E. is correct
Glucose is absorbed by a Tm-limited transport mechanism.

**Use the following to answer questions 11 and 12:**
A. phosphate
B. protein
C. water
D. sodium
E. bicarbonate

4.011 Elevation in circulating aldosterone results in a decrease in excretion.

D. is correct
Aldosterone increases sodium reabsorption and promotes potassium secretion.

4.012 Tubular transport maximum (Tm) is reduced by parathyroid hormone.

A. is correct
Parathyroid hormone (PTH) promotes phosphate excretion by the kidney.

**Use the following to answer questions 13 and 14:**

    A. proximal convoluted tubule
    B. thin, descending limb of the loop of Henle
    C. thick, ascending limb of the loop of Henle
    D. distal convoluted tubule
    E. collecting duct

4.013 Tubular fluid leaving the nephron segment is always more dilute than systemic plasma.

C. is correct
The thick ascending limb of the loop of Henle is impermeable to water and results in a dilute tubular fluid.

4.014 Water permeability of entire segment is regulated by antidiuretic hormone.

E. is correct
Antidiuretic hormone (ADH) increases the permeability of distal convoluted tubule and collecting duct to water.

4.015 Plasma volume in a subject is 2.8 liters. If his hematomcrit is 33%, blood volume would be closest to:
    A. 3.8 liters
    B. 4.2 liters
    C. 5.0 liters
    D. 7.5 liters
    E. 8.1 liters

B. is correct

$$BV = PV \frac{100}{100 - Hct} = 2.8 \times 1.5 = 4.2 l$$

**Use the following to answer question 16:**

Radioactive chloride is administered to a subject and the following data were obtained following equilibration:
Amount of radioactive chloride administered: $9 \times 10^6$ dpm
Radioactive chloride in plasma: 500 dpm/ml
Plasma chloride concentration: 102 mEq/liter

4.016 If radioactive chloride was not lost from the body during the equilibration period, the amount of exchangeable chloride in this subject would be closest to:
    A. 1,100 mEq
    B. 1,800 mEq
    C. 2,100 mEq
    D. 2,800 mEq
    E. 3,400 mEq

B. is correct

$$Cl_E = \frac{^{24}Cl(9 \times 10^6 dpm)}{SA \text{ of chloride, cpm/mEq}}$$

$$SA = \frac{500,000 \text{ dpm/l}}{102 \text{ meq/l}}$$

$$Cl_E = \frac{9 \times 10^6 \text{ dpm}}{4902 \text{ dpm/mEq}} = 1836 \text{ mEq}$$

**Use the following to answer question 17:**

The following data are from a normal subject:
Total body water:  40 liters
Plasma osmolality:  300 mOsm/Kg water
Extracellular fluid volume:  13 liters
Intracellular fluid volume:  27 liters

4.017  If this subject drinks 750 ml water, plasma osmolality following equilibration would be closest to:
A.  280 mOsm/kg water
B.  285 mOsm/kg water
C.  290 mOsm/kg water
D.  295 mOsm/kg water
E.  315 mOsm/kg water

D. is correct
Solutes in ECF (13 x 300 = 3900 mOsm)
DCF (27 x 300 = 8100 mOsm)

$$Posm = \frac{12000 \text{ mOsm}}{40} \div 40 = 300 \text{ mOsm}$$

**Use the following to answer questions 18 and 19:**

The following data are from a normal subject:
Plasma concentration:
creatinine:  1.1 mg/100 ml
glucose:  220 mg/100ml
Urinary concentration:
creatinine:  1.51 mg/100 ml
glucose:  60 mg/ml
Urine flow:  0.8 ml/min

4.018  Glomerular filtration rate is closest to:
A.  75 ml/min
B.  100 ml/min
C.  110 ml/min
D.  125 ml/min
E.  140 ml/min

C. is correct

$$GFR = \frac{U_{cr}V}{Pcr} = \frac{1.51 \text{ mg/min x 80 ml/min}}{1.1 \text{ mg/min}}$$

= 109. 8 or 110 ml/min

4.019  If glomerular filtration rate in this subject was 90 ml/min, the rate of tubular reabsorption of glucose would be closest to:
A.  100 mg/min
B.  125 mg/min
C.  150 mg/min
D.  220 mg/min
E.  375 mg/min

C. is correct
GFR x $P_G$-$T_G$ = VGV-TG
90 x 2.2 - X = 60
X = 198 - 60 = 138 mg/min
Answer is closest to 150.

4.020  A substance that distributes in the extracellular fluid and whose clearance estimates glomerular filtration rate:
A.  tritiated water
B.  radioactive sodium
C.  inulin
D.  Evans Blue (T-1824)
E.  antipyrine

C. is correct
Inulin is neither reabsorbed or secreted and therefore inulin clearance estimates GFR.

4.021  Fluid compartment that is smaller than the other four:
  A. interstitial fluid
  B. extracellular fluid
  C. plasma
  D. transcellular fluid
  E. intracellular fluid

D. is correct
Transcellular fluid composed of fluids formed by epithelial tissues (e. g. gastrointestinal secretins, CSF, etc) accounts for about 1. 5% of the total.

4.022  Causes osmolality of body fluids to increase, extracellular fluid volume to increase, and intracellular volume to decrease:
  A. water deprivation
  B. loss of body sodium without loss of water
  C. ingestion of distilled water
  D. intravenous administration of isosmotic glucose
  E. ingestion of NaCl tablets with a negligible volume of water

E. is correct
Ingestion of sodium chloride tablets will increase the number of particles in body fluids and extracellular fluid volume and decrease the intracellular volume.

4.023  Increases in both glomerular filtration rate and renal blood flow are caused by:
  A. decrease in the surface area available for filtration
  B. constriction of the efferent arterioles in the kidney
  C. increase in renal sympathetic nerve activity
  D. increase in colloid oncotic pressure of arterial plasma
  E. dilation of the afferent arterioles in the kidney

E. is correct
Only dilation of the afferent arterioles results in increased renal blood flow and GFR.

4.024  All of the following are transported across the tubule by an active, Tm-limited mechanism:
  A. glucose
  B. p-aminohippuric acid (PAH)
  C. phosphate
  D. bicarbonate
  E. amino acid

D. is correct
Only bicarbonate is not transported by an active Tm-limited mechanism.

4.025  A reduction in glomerular filtration rate causes an increase in concentration in blood or plasma of:
  A. chloride
  B. creatinine
  C. sodium
  D. potassium
  E. protein

B. is correct
Creatinine is an endogenous metabolite formed from phosphorylcreatinine. It is mainly excreted by filtration, thus a reduction in GFR would cause an increased concentration in plasma.

4.026  The generation of an osmotic gradient between cortical and medullary interstitial fluids results from active reabsorption of sodium or chloride in the:
  A. proximal convoluted tubule
  B. descending limb of the loop of Henle
  C. thick, ascending limb of the loop of Henle
  D. distal convoluted tubule
  E. collecting duct

C. is correct
The thick ascending loop of Henle reabsorbs about 25% of the filtered $Na^+$. Sodium and two chloride ions are co-transported across the luminal membrane.

4.027 Plasma concentration of a substance, X, that is freely filterable is 2 mg/ml. If renal plasma flow is 600 ml/min and glomerular filtration rate is 120 ml/min the rate that X is delivered to the peritubular capillary system is closest to:
  A. 240 mg/min
  B. 450 mg/min
  C. 650 mg/min
  D. 950 mg/min
  E. 1200 mg/min

D. is correct
The amount of X entering the peritubular system is Px(RPF) - px(GFR) or 1200 mg/min - 240 mg/min = 960 mg/min.

4.028 Passive secretion of potassium in the distal nephron decreases in response to an increase in:
  A. negative potential in the luminal compartment
  B. delivery of fluid to the distal nephron
  C. activity of $Na^+K^+$-ATPase
  D. secretion of hydrogen ion
  E. potassium concentration within tubule cells

D. is correct
Potassium and $H^+$ are exchanged for sodium in the distal collecting tubules and collecting duct. Thus, if secretion of $H^+$ is increased less passive secretion of potassium occurs.

4.029 Which of the following would tend to stimulate secretion of antidiuretic hormone:
  A. intravenous infusion of isotonic NaCl
  B. reduction in plasma osmolality
  C. ethanol
  D. decrease in firing frequency of baroreceptors
  E. decrease in circulating levels of renin

D. is correct
A decrease in firing frequency of mechanoreceptors resulting from hypotension or hypovolemia enhances ADH secretion.

4.030 Depletion of body potassium content would most likely result from an increase in circulating levels of:
  A. antidiuretic hormone
  B. aldosterone
  C. atrial natriuretic factor
  D. parathormone
  E. renin

B. is correct
Aldosterone increases sodium reabsorption and promotes passive secretion of potassium.

4.031 An enzyme that causes renal formation of ammonia:
  A. glutaminase
  B. ATPase
  C. adenylate cyclase
  D. carbonic anhydrase
  E. renin

A. is correct
The rate of synthesis of $NH_3$ by tubular cells is influenced by the level of glutaminase activity. During chronic acidosis there is an adaptive increase in glutaminase activity.

4.032 In a normal individual, over 90% of the hydrogen ions secreted by the nephrons are utilized in the:
  A. formation and excretion of titratable acid
  B. excretion of ammonium ions
  C. reabsorption of filtered bicarbonate
  D. excretion of free hydrogen ions
  E. titration of beta hydroxybutyrate to its weak acid

C. is correct
Most hydrogen ions excreted by the tubule are not excreted but are used to reabsorb filtered bicarbonate.

4.033 All of the following would occur following expansion of blood volume EXCEPT:
 A. increase in blood pressure
 B. decrease in secretion of antidiuretic hormone
 C. decrease in circulating levels of renin
 D. increase in renal blood flow
 E. arousal of thirst

E. is correct
Thirst arousal would be least likely to occur after expansion of blood volume.

4.034 All of the following forces impact on the glomerular filtration rate (GFR) EXCEPT:
 A. hydrostatic pressure in the glomerular capillary
 B. plasma colloid oncotic pressure
 C. hydrostatic pressure in Bowman's capsule
 D. total osmotic pressure of the glomerular filtrate
 E. colloid osmotic pressure of one filtrate

D. is correct
Total osmotic pressure is the osmotic pressure that results when the membrane is impermeable to all of the substances in the filtering solution. Therefore, total osmotic pressure plays no role.

4.035 The clearance of inulin will be:
 A. increased in complete bilateral ureteral obstruction
 B. positively correlated with the rate of infusion of inulin
 C. positively correlated with the GFR
 D. dependent upon active transport mechanisms
 E. decreased when the renal blood flow increases

C. is correct
The clearance of inulin is a measure of glomerular filtration rate.

$$C_{IN} = \frac{U_{IN} \times V}{P_{IN}}$$

4.036 Forces opposing glomerular filtration include which of the following?
 A. colloid osmotic pressure of the filtrate
 B. hydrostatic pressure in Bowman's capsule
 C. glomerular capillary pressure
 D. crystalloid osmotic pressure of the final urine
 E. increase in renal blood flow

B. is correct
Hydrostatic pressure in Bowman's capsule opposes ultrafiltration by the glomerulus.

4.037 Osmotic diuretics exhibit all of the following characteristics EXCEPT:
 A. fully filterable by the glomerulus
 B. metabolically altered to an active form in the proximal tubule
 C. must increase the number of filtrate particles
 D. must be incompletely reabsorbed in the nephron
 E. must not induce adverse changes in the lining of the proximal tubule

B. is correct
The osmotic diuretic must be able to be filtered by the glomerulus, the number of particles in the filtrate determine the osmolality and extra particles must remain in the filtrate to be osmotically active.

4.038 Renal plasma flow is 850 ml/min with a glucose concentration of 90 mg/100 ml of plasma and the GFR is 125 ml/min, the filtered load of glucose would be closest to:
 A. 190 mg/min
 B. 113 mg/min
 C. 468 mg/min
 D. 18 mg/min
 E. 225 mg/min

B. is correct
To determine the filtered load, multiply GFR x concentration of substance in plasma. 125 x .9 = 112. 5 mg/min.

4.039 Net secretion by the kidney into the tubular lumen can be determined when which of the following occurs?
- A. it has a greater concentration in the urine than plasma
- B. the concentration increases in the lumen as it passes through the tubule
- C. the urinary concentration increases with increases in plasma concentration
- D. its clearance is greater than inulin
- E. the clearance of inulin exceeds the clearance of PAH

D. is correct
Any time clearance of a substance is greater than inulin then secretion is indicated since inulin is not secreted or reabsorbed.

4.040 All of the following alter glomerular filtration rate EXCEPT:
- A. sympathetic stimulation
- B. decreased vascular tone in the kidney
- C. change in mean arterial blood pressure from 100 to 120 mmHg
- D. vasoconstriction of efferent arterioles
- E. epinephrine

C. is correct
GFR remains unchanged even with larger changes in arterial blood pressure. Filtration across the glomerular basement membrane depends on the difference between glomerular hydrostatic pressure and the sum of Bowman's capsule hydrostatic pressure and the plasma protein osmotic pressure.

4.041 In the normal young adult at rest:
- A. renal blood flow is approximately 10% of cardiac output
- B. the percentage of filtered inulin that is excreted is about 20
- C. renal blood flow is greater in the cortex than the medulla.
- D. extraction of PAH may be 85 - 90%
- E. secretion can account for over 75% of the PAH that is excreted

C. is correct
Renal blood flow is 20 - 25% of the cardiac output and is greater in the renal cortex than medulla.

4.042 Which of the following is associated with an increase in the filtration fraction?
- A. increased ureteral pressure
- B. increased plasma protein concentration
- C. increased efferent arteriolar resistance
- D. decreased glomerular capillary pressure
- E. decreased glomerular filtration area

C. is correct
Constriction of efferent arterioles results in an increase in glomerular filtration rate and glomerular pressure.

4.043 All of the following are correct regarding aldosterone EXCEPT:
- A. it is released from the adrenal medulla
- B. its secretion is directly stimulated by angiotensin II
- C. it causes net retention of body sodium
- D. it is a steroid
- E. it promotes $K^+$ secretion in the distal convoluted tubule

A. is correct
Aldosterone is synthesized by cells of the zona glomerulosa of the adrenal cortex and stimulates sodium retention and potassium secretion.

4.044 After ingestion of potassium, which of the following is responsible for the majority of its appearance in urine:
- A. GFR
- B. secretion in the proximal tubule
- C. GFR and proximal tubular secretion
- D. secretion in the distal nephron
- E. secretion in the proximal tubule and the loop of Henle

D. is correct
Most potassium appears in the urine as a consequence of secretion in the distal tubule.

4.045 Determination of resistance to blood flow through the kidney can be accomplished by which of the following?
- A. a manometer connected to a cannula inserted in a resistance vessel of the kidney
- B. measuring the pressure difference between the renal artery and vein
- C. dividing the rate of renal blood flow by the A/V pressure difference
- D. dividing the A/V difference by the rate of blood flow
- E. measuring the rate of blood flow

D. is correct
Resistance to renal blood flow can be determined by dividing pressure by the rate of blood flow.

4.046 An increase in plasma concentration of inulin results in which of the following regarding the clearance of inulin?
- A. increase
- B. decrease
- C. no change
- D. increase first then decrease
- E. decrease first then increase

C. is correct
As plasma concentration of inulin increases, the clearance of inulin remains unchanged. Inulin is neither reabsorbed nor secreted and provides a good estimate for GFR.

4.047 Which of the following would occur to clearance of PAH when the plasma concentration of PAH is increased from 5 to 50 mg/100cc?
- A. increase
- B. decrease
- C. no change
- D. increase first, then decrease
- E. decrease first, then increase

D. is correct
As the plasma concentration of PAH increases, the clearance first increases then decreases. This illustrates what occurs when the concentration of PAH is increased so that the transport maximum ($T_m$) for PAH secretion is exceeded.

4.048 A decrease in the urine plasma rates (U/P) of inulin when the GFR is constant would indicate:
- A. inulin clearance has decreased
- B. free water clearance has decreased
- C. osmolar clearance has decreased
- D. plasma inulin concentration has decreased
- E. urine formation has increased

E. is correct
Urine flow has increased. The clearance of any substance (S) is calculated as follows:

$$\text{Clearance}_s = \frac{\text{Urine concentration}_s}{\text{Plasma concentration}_s} \times \text{urine flow rate}$$

4.049 When a drug that blocks secretion of PAH is administered to a patient, which of the following compounds will have the same clearance as PAH under this condition?
- A. water
- B. sodium
- C. urea
- D. bicarbonate
- E. inulin

E. is correct
PAH is freely filtered at the glomerulus and is also actively secreted into the nephron. When PAH secretion is blocked, PAH is only filtered. Therefore, the clearance of inulin and PAH would be the same under these conditions since both are freely filtered.

4.050 When a normal subject receives an intravenous infusion of 250 ml of a 5% sodium chloride solution, one would expect:
- A. urinary solute excretion to be decreased
- B. ECF volume to be decreased
- C. urine volume to be increased
- D. ICF volume osmolality to be decreased
- E. sodium excretion to be decreased

C. is correct
A 5% NaCl solution is hypertonic and more water would move from the cells to the lumen, thereby increasing urine volume.

4.051 Renin release is stimulated by:
- A. an increase in renal arterial pressure
- B. a decrease in delivery of NaCl to the macula densa
- C. an increase in circulating ADH
- D. a decrease in renal sympathetic nerve activity
- E. an increase in circulating angiotension II levels

B. is correct
Renin secretion is increased by stimuli that decrease ECF volume, and blood pressure or that increase sympathetic output. It appears that renin secretion is inversely proportionate to the rate of NaCl across the macula densa.

4.052 Which of the following tissues contains less than 50% water?
- A. skeletal muscle
- B. bone
- C. intestine
- D. liver
- E. blood

B. is correct
Bone contains about 20% water.

4.053 Which of the following indications are used to measure intestinal fluid volume in humans?
- A. antipyrine
- B. $Cr^{54}$ - labeled erythrocytes
- C. Evan's Blue
- D. $D_{20}$
- E. PAH

C. is correct
The volume of ISF = ECF - PV. Thus, ISFV is estimated by subtracting the volume of distribution of Evan's Blue (T-1824) from the volume of inulin.

4.054 Which of the following has a higher concentration (mEq/liter water) in ISF than plasma?
- A. protein
- B. chloride
- C. potassium
- D. bicarbonate
- E. sodium

B. is correct
The concentration of diffusable chloride in ISF is greater than in plasma.

4.055 Which of the following contains the largest pool of body sodium?
- A. transcellular fluid
- B. intracellular fluid
- C. bone
- D. plasma
- E. dense connective tissue and cartilage

C. is correct
Bone contains the largest amount of body sodium most of which is non-exchangeable.

4.056  Which of the following correctly define renal threshold for glucose?
- A. rate of filtration of glucose
- B. plasma glucose concentrations when Tm for glucose is reached
- C. the degree of splay in the glucose filtration curve
- D. plasma glucose concentration when glucose first appears in urine
- E. the amounts of filtered urea that is reabsorbed

D. is correct
Renal threshold for glucose refers to plasma glucose concentration when glucose first appears in urine.

4.057  All of the following are actions of arterial natriuretic peptide EXCEPT:
- A. inhibition of renin secretion
- B. decreased blood pressure
- C. increased GFR
- D. inhibited ADH secretion
- E. increased renal sodium retention

C. is correct
ANP is a peptide hormone that causes sodium excretion to increase. ANP decreases blood pressure and increases GFR and inhibits sodium reabsorption.

4.058  Aldosterone:
- A. is released from the adrenal medulla
- B. site of action is the luminal membrane of the nephron
- C. secretion is stimulated by angiotensin II
- D. causes net retention of body potassium
- E. promotes enhanced renal sodium excretion

C. is correct
Aldosterone is released from the adrenal cortex in response to angiotensin II and increases sodium retention and potassium excretion.

4.059  ADH release may be stimulated by a decrease in:
- A. blood volume
- B. release of renin
- C. plasma osmolality
- D. thirst
- E. sodium intake

A. is correct
Only a decrease in blood volume will increase ADH, all the rest are associated with a decrease in ADH release.

4.060  A subject drinks over a liter of isotonic NaCl rapidly. This will result in:
- A. a decrease in $P_{osm}$:
- B. an increase in blood volume
- C. a decrease in GFR
- D. a water diuresis
- E. a decrease in ANP secretion

B. is correct
An increase in sodium intake causes an increase in plasma osmolality which stimulates ADH secretion and arouses thirst.

4.061  Systemic angiotensin II may alter the firing frequency of neurons in the paraventricular nucleus by:
- A. entering the brain by diffusion through the blood-brain barrier
- B. lowering blood pressure
- C. interaction with the subfornical organ
- D. acting on the brain osmoreceptors
- E. stimulating the pineal gland

C. is correct
Among the several actions of angiotensin II, via the subfornical organ to promote ADH release and possibly thirst.

4.062 At normal arterial pH, the quantitatively important buffer anions in blood include:
A. phosphate
B. chloride
C. sulfate
D. bicarbonate
E. hydrogen ion

D. is correct
The carbonic acid-bicarbonate system is a major system along with plasma protein and hemoglobin.

4.063 An arterial blood sample has a buffer base concentration of 40 mE/l blood. Base excess for this sample equals:
A. +12 mE/l
B. -10 mEq/l
C. 0. 0
D. -8 mEq/l
E. e + 10 mEq/l

D. is correct
Buffer base is the sum of the concentrations of conjugate bases of buffers in arterial blood. The sum of these is 48 mEq/l BB = $BB_{obs}$ – 48 mEq/l blood. BB = 40 – 48 Eq/L = –8 mEq/l

4.064 The bicarbonate-carbonic acid system is an important buffer system in the body because:
A. it has a pK value that makes it an effective chemical buffer
B. the weak acid form of the buffer can be adjusted by the respiratory system
C. the conjugate base of the buffer system is regulated by respiration
D. it is an important buffer anion that is normally excreted by the kidneys
E. the acid form of the buffer is regulated by renal mechanisms

B. is correct
The system is one of the most effective buffer systems in the body because the amount of dissolved $CO_2$ is controlled by respiration.

**Use the following to answer questions 65-68.**

A. asthma
B. going from sea-level to 14,000 ft altitude
C. ingestion of $NH_4Cl$
D. excessive vomiting
E. diabetes mellitus

4.065 Metabolic alkalosis:

D. is correct
Metabolic alkalosis is caused by addition of base or loss of acid. Prolonged vomiting results in loss of $H^+$.

4.066 Respiratory acidosis:

A. is correct
Respiratory acidosis is a primary increase in $P_aCO_2$ due to a decrease in alveolar ventilation rate.

4.067 A primary (uncompensated) increase in plasma [$HCO_3$]:

D. is correct
Excessive vomiting leads to severe gastric acid loss which results in an increase in plasma bicarbonate.

4.068  A primary (uncompensated) decrease in $P_aCO_2$:

B. is correct
Going from sea-level to 14,000 ft. altitude results in a reduction in $P_aCO_2$ through increased respiration by stimulation of the respiratory center.

4.069  The reabsorption of filtered $HCO_3^-$ in the proximal convoluted tubule would decrease if:
   A.  carbonic anhydrase activity is inhibited
   B.  partial pressure of $CO_2$ increases
   C.  chloride concentration of filtrate decreases
   D.  body potassium stores are depleted
   E.  body sodium stores are depleted

A. is correct
Carbonic anhydrase catalyzes the formation of $H_2CO_3$ and drugs that inhibit carbonic anhydrase decreases the secretion of acid.

4.070  The rate of excretion of $NH_4^+$ increases with:
   A.  chronic hyperventilation (high altitude)
   B.  decrease in renal glutaminase activity
   C.  chronic ketoacidosis
   D.  increase in pH of tubular fluid

C. is correct
The rate of excretion of $H^+$ in the form of $NH_4^+$ depends upon the rate of synthesis and secretion of $NH_3$, the pH of tubular fluid and the rate of urine flow. During chronic acidosis there is an increase in glutaminase activity increasing the amount of $NH_4^+$ excreted.

4.071  Plasma $HCO_3^-$ concentration increases with:
   A.  metabolic acidosis
   B.  metabolic alkalosis
   C.  respiratory acidosis
   D.  respiratory alkalosis
   E.  starvation

B. is correct
Metabolic alkalosis is an acid base disturbance characterized by increased plasma bicarbonate and plasma pH.

**Use the table below to answer questions 72-75.**

| Sample | pH | [$HCO_3^-$] mEq/l | $PCO_2$ mmHg | $BB_b$ mEq/l |
|--------|------|------|------|------|
| normal | 7. 4 | 24 | 40 | 48 |
| A | 7. 2 | 10 | 27 | 30 |
| B | 7. 3 | 35 | 70 | 53 |
| C | 7. 5 | 32 | 40 | 60 |
| D | 7. 5 | 22 | 28 | 48 |

$BB_b$ = Buffer base value of blood

4.072  Uncompensated metabolic alkalosis

C. is correct
Metabolic alkalosis is characterized by increased plasma bicarbonate and alkalemia (high blood plasma pH).

4.073  Partially compensated respiratory acidosis

B. is correct
Respiratory acidosis is characterized by $PCO_2$ above 50 mmHg and acidemia (low blood plasma pH).

4.074  Uncompensated respiratory alkalosis

D. is correct
Respiratory alkalosis is characterized by decreased plasma bicarbonate and alkalemia.

4.075 Partially compensated diabetic ketoacidosis

A. is correct
Metabolic acidemia is characterized by decreased plasma bicarbonate and acidemia due to acid overproduction and is often the result of diabetic ketoacidosis.

4.076 Release of antidiuretic hormone is stimulated by:
    A. ethanol
    B. decrease in plasma osmolality
    C. isosmotic expansion of extracellular fluid volume
    D. hemorrhagic hypotension
    E. water consumption

D. is correct
ADH release is stimulated by water deficit, increased plasma osmolality, decreased atrial pressure and volume deficit.

4.077 Intravenous injection of a large dose of epinephrine in an anesthetized dog causes a decrease in urine flow.
    A. an increase in glomerular filtration rate
    B. an increase in plasma oncotic pressure
    C. an increase in hydrostatic pressure in Bowman's capsule.
    D. a decrease in efferent arteriolar resistance
    E. a decrease in glomerular hydrostatic pressure

E. is correct
Catecholamines constrict renal blood flow and decrease glomerular hydrostatic pressure.

4.078 All of the following are associated with increase in renal blood flow EXCEPT:
    A. increased rate of filtration of sodium
    B. increased renal oxygen consumption
    C. increased tubular reabsorption of sodium
    D. increased renal extraction of oxygen
    E. increased renal excretion of sodium

D. is correct
The extraction of oxygen increases as blood flow decreases to keep oxygen consumption constant.

**Choose the single best answer:**
**Use the following to answer questions 79-81:**

    A. p-aminohippuric acid
    B. ammonia
    C. salicylate
    D. hydrogen ion
    E. potassium ion

4.079 Endogenously produced substance that is passively secreted by a process termed "non-ionic diffusion":

B. is correct

4.080 Passive secretion in the distal convoluted tubule is inversely related to rate of hydrogen ion secretion:

E. is correct

4.081 Clearance is equal to effective renal plasma flow:

A. is correct

**Use the following to answer questions 82-84:**

    A. antidiuretic hormone
    B. angiotensin II
    C. aldosterone
    D. atrial natriuretic factor
    E. renin

4.082 Action of kidneys results in an increase in renal potassium excretion

C. is correct
Aldosterone increases sodium retention and potassium excretion.

4.083 Arouses thirst in experimental animals and possibly man

B. is correct
Angiotensin II is believed to stimulate the subfornical organ and OVLT to increase water intake.

4.084 Genetic absence of hormone receptor results in nephrogenic diabetes insipidus

A. is correct
Diabetes insipidus is due to a deficiency of ADH.

**EXTENDED MATCHING:**

**Use the following to answer questions 85-89:**

    A. chloride
    B. ammonium
    C. sulfate
    D. bicarbonate
    E. phosphate
    F. glucose
    G. para-aminohippuric acid
    H. magnesium
    I. calcium

4.085 Quantitatively important in the formation and excretion of titratable acid by the kidneys

E. is correct

4.086 Excretion increases daily during the first five to seven (5-7) days of chronic metabolic acidosis due to an adaptive enzyme system

B. is correct

4.087 Not excreted by the kidneys when urine is acid.

D. is correct

4.088 Clearance increases as plasma concentration is progressively elevated

F. is correct

4.089 Tubular reabsorption can be suppressed by parathyroid hormone

E. is correct

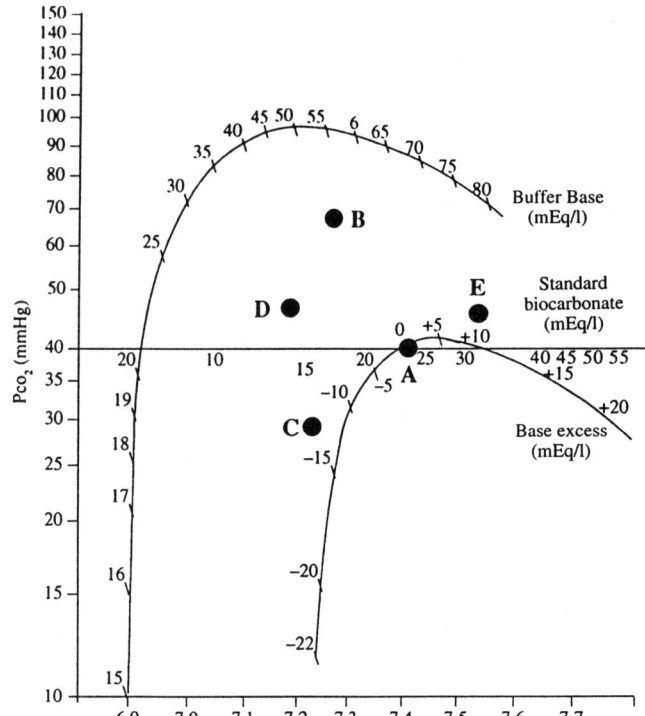

**Use the above nomogram to answer questions 90 and 91. Which of the lettered points on the nomogram best fit the conditions described below?**

4.090  Partially compensated metabolic acidosis          B. is correct

4.091  Uncompensated respiratory acidosis               D. is correct

**Extended Matching**

Use the following to answer questions 92 and 93.
Match the situation with its appropriate acid-base disturbance. Each letter may be used once, more than once, or not at all.

   A.  salicylate ingestion
   B.  pyloric stenosis
   C.  Crohn's disease
   D.  methanol ingestion
   E.  furosemide
   F.  opiate ingestion
   G.  multiple myeloma
   H.  cystic fibrosis
   I.  hepatic failure

4.092  Respiratory acidosis                              F. is correct

4.093  Respiratory alkalosis                             I. is correct

**Use the following to answer questions 94-96:**

Select the most appropriate renal concept or function for each of the descriptions below. Each may be used once, more than once or not at all.

    A. active transport of sodium
    B. clearance of inulin
    C. excretion rate
    D. filtration fraction
    E. glomerular filtration rate
    F. hydrostatic pressure in the glomerulus
    G. oncotic pressure in the glomerulus
    H. reabsorption rate
    I. renal blood flow
    J. renal plasma flow
    K. secretion rate
    L. tubular maximum

4.094 The rate of oxygen consumption of the kidney correlates best with this renal function
    A. is correct

4.095 Mathematically, the clearance of PAH divided by 90% is equivalent to this important renal function
    J. is correct

4.096 Autoregulation in the kidney results in a relatively constant renal plasma flow. Which of the concepts listed is most constant due to this homeostatic regulation of renal plasma flow?
    F. is correct

**Use the following to answer questions 97 and 98.**

    A. renal vein
    B. vasa recta
    C. inferior vena cava
    D. efferent arteriole
    E. peritubular capillary
    F. afferent arteriole
    G. renal artery
    H. aorta
    I. glomerulus

4.097 A countercurrent, capillary system found in the renal medulla
    B. is correct

4.098 Resistant increases in response to an increase in arterial pressure, thus autoregulating glomerular filtration rate
    F. is correct

**Use the following to answer questions 99 and 100**

    A. phosphate
    B. chloride
    C. inulin
    D. water
    E. urea
    F. sodium
    G. glucose
    H. albumin
    I. creatinine

4.099  An endogenous substance whose clearance is used clinically to measure glomerular filtration rate

    I. is correct

4.100  Co-transported together with sodium and potassium from luminal fluid into the cell interior in the thick ascending limb of the loop of Henle

    B. is correct

# SECTION 5:   GASTROINTESTINAL PHYSIOLOGY

5.001  The caloric equivalent of dietary lipid is:
   A.  0.5 cal/g (2.1 joules/g)
   B.  0.9 cal/g (3.8 joules/g)
   C.  4.0 cal/g (16.4 joules/g)
   D.  7.3 cal/moles (30.7 joules/mole)

B. is correct
The caloric equivalent for fat is 0.9 cal/g, while the caloric equivalent for carbohydrate and protein is around 4.0 cal/g.

5.002  Chymotrypsinogen is converted to chymotrypsin by the action of:
   A.  pepsin
   B.  trypsin
   C.  enterokinase
   D.  acid
   E.  lysosomal enzymes

B. is correct
Trypsin converts all proteolytic proenzymes to their active forms. Enterokinase from the intestinal mucosa activates trypsinogen to trypsin.

5.003  Fatty acids are most potent in inhibiting gastric motility:
   A.  placed in the mouth
   B.  placed in the stomach
   C.  placed in the duodenum
   D.  placed in the ileum
   E.  injected intravenously

C. is correct
Foodstuffs in the small intestine cause a number of neural and hormonal responses to slow gastric motility.

5.004  A correct statement containing short-chain fatty acids include(s):
   A.  they can be absorbed by the small intestine without undergoing re-esterification
   B.  they must be incorporated into chylomicrons for efficient absorption
   C.  they require pancreatic lipase digestion
   D.  they are actively transported from the lumen by carriers located in the brush border of intestinal cells
   E.  none of the above

A. is correct
Short chain fatty acids do not require intralumenal hydrolysis and are extruded from intestinal mucosal cells inthe form of free fatty acids into the portal blood.

5.005  The volume output of pancreatic secretion is due almost entirely to the stimulating action of:
   A.  sympathetic secretory fibers
   B.  cholecystokinin
   C.  pancreozymin
   D.  secretin
   E.  none of the above

D. is correct
Secretin is primarily responsible for the watery bicarbonate secretion of the pancreas.

5.006  A patient deficient in lactase has a primary abnormality that involves the:
   A.  oxyntic cells of the stomach
   B.  pancreas
   C.  small intestinal epithelium
   D.  salivary glands
   E.  liver

C. is correct
The brush cell border of the small intestine contains many enzymes for hydrolysis of disaccharides, peptides and nucleic acids.

**Use the following to answer question 7.**

1   Triglyceride re-esterification
2   Micelle
3   Lactea l
4   CCK
5   Chylomicron

5.007  Normally, the sequence in which the items listed above are involved in lipid absorption is:
   A.  1-2-3-4-5
   B.  2-4-3-1-5
   C.  4-5-2-1-3
   D.  4-2-1-5-3
   E.  4-1-5-2-3

D. is correct
Absorption of fat involves the formation of micelles. Pancreatic lipase acts on micelles releasing fatty acids and 2-monoglycerides that are taken up into mucosal cells. Fatty acids are then incorporated into phospholipids and triglycerides and released into lymph as chylomicrons.

5.008  Obstruction of the bile duct would result in:
   A.  a decreased excretion of bilirubin glucuronides in the urine
   B.  increased content of lipids in feces
   C.  decreased conjugated bile pigments in the serum
   D.  increased enterohepatic circulation of bile acids or salts
   E.  all of the above

B. is correct
Bile salts are necessary for adequate digestion and absorption of fat. Obstruction of the bile ducts results in a spillover of conjugated bilirubin in plasma and its eventual excretion in urine. With bile duct obstruction less bile salts would undergo enterohepatic circulation.

5.009  Low gastric pH:
   A.  inhibits release of gastrin
   B.  inhibits release of histamine
   C.  inhibits peptic activity
   D.  reduces irritation caused by aspirin
   E.  none of the above

A. is correct
Intragastric pH< 2.0 will inhibit further release of gastrin. This phenomenon is thought to be mediated by somatostatin.

5.010  During ingestion of a meal containing protein:
   A.  antral cells are stimulated by peptides to release gastrin.
   B.  vagal discharge stimulates both gastrin release and parietal cell acid secretion
   C.  the chief cells release pepsinogen
   D.  antral gastrin secretion is inhibited by the secreted acid
   E.  all of the above

E. is correct
Gastrin is released by vagal impulses and protein digestive products that bathe the antral mucosa. Gastrin stimulates release of HCl by oxyntic cells and pepsinogen by chief cells. Further gastrin release is inhibited by excess gastric acidity.

5.011  The absorption of iron by the small intestine:
   A.  is greatest in the ileum
   B.  increases to maximal levels within a few hours following hemorrhage
   C.  shows a preference for ferrous ion
   D.  is enhanced by the alkaline secretion of the pancreas
   E.  all of the above

C. is correct
Iron is primarily absorbed in the duodenum in the ferrous state. The rate of absorption increases after hemorrhage but not for three days due to migration of iron deficient cells to the tips of the villi three to four days later.

5.012 The major site for vitamin $B_{12}$ absorption:
  A. stomach
  B. duodenum
  C. jejunum
  D. ileum
  E. descending colon

D. is correct
Although most vitamins are absorbed in the upper small intestine, vitamin $B_{12}$ is selectively absorbed in the ileum.

5.013 The most potent stimulus for the release of secretin is:
  A. fat
  B. carbohydrate
  C. protein
  D. acid
  E. $Ca^{++}$

D. is correct
Acidification of the small intestine causes release of the hormone secretin.

5.014 Gastrin:
  A. shares the same C-terminal amino acids with cholecystokinin
  B. its action on gastric acid secretion is antagonized by secretion
  C. potentiates the action of acetylcholine and histamine on gastric parietal cells
  D. its secretion is inhibited by low gastric pH
  E. all of the above

E. is correct
Gastrin shares the same terminal amino acids with CCK and the action of acetylcholine and histamine on acid production. Gastrin stimulation of acid secretion is inhibited by secretin. In addition, gastrin has a trophic effect on the gastric mucosa.

5.015 Which of the following is(are) true for the colon?
  A. sodium is actively absorbed
  B. chloride and bicarbonate are transported in opposite directions by an ion exchange mechanism
  C. aldosterone promotes the release of potassium into the lumen
  D. water is absorbed by osmosis
  E. all of the above

E. is correct
The colonic mucosa actively reabsorbs sodium and promotes potassium secretion under the influence of aldosterone. Chloride is also exchanged for bicarbonate in the large intestine.

5.016 Micelle formation is a prerequisite to the efficient absorption of the following EXCEPT:
  A. vitamin A
  B. cholesterol
  C. linolenic acid
  D. vitamin K
  E. glycerol

E. is correct
Micelle formation is necessary for absorption of fat soluble vitamins (A, D, E and K) as well as cholesterol and phospholipids.

5.017 Which of the following are correct for vomiting?
  A. contraction of abdominal muscles
  B. relaxation of the body of the stomach
  C. contraction of the antral-pyloric area
  D. relaxation of the lower esophageal sphincter
  E. all of the above

E. is correct
Vomiting is characterized by strong contractions of abdominal muscles, relaxation of the lower esophageal sphincter, and body of the stomach, and spasms of the gastric antrum to force material up toward the mouth.

5.018 Which of the following are all released as proenzymes?
  A. chymotrypsin, amylase, lipase
  B. chymotrypsin, secretin, elastase
  C. trypsin, amylase, carboxypeptidase A
  D. trypsin, elastase, carboxypeptidase B
  E. lingual lipase, salivary amylase, gastric lipase

D. is correct
Only proteolytic enzymes are secreted as inactive proenzymes to prevent pancreatic autodigestion.

**Use the following to answer questions 19-20**

    A. rate of gastric emptying of a solid meal
    B. rate of duodenal motor activity
    C. rate of gastric slow waves
    D. rate of the migrating myoelectric complex (MMC)
    E. none of the above

5.019  Motor activity of the antrum

A. is correct
Gastric emptying is due to the antral pump activity.

5.020  Fundic pacemaker cells

C. is correct
Gastric slow waves or B. E. R. are coordinated by pacemaker cells located in the gastric fundus.

5.021  Motilin

D. is correct
Plasma levels of motilin rise with the generation of MMCS.

**Use the following information to answer questions 22-24.**

    A.    removal of the fundus (body) of the stomach
    B.    removal of the ileum
    C.    removal of antrum of the stomach
    D.    none of the above

5.022  Abnormally high serum concentrations of gastrin after a meal

A. is correct
Gastrin is produced by the gastric antrum and stimulates acid production by acid producing cells in the fundus.

5.023  Pernicious anemia

B. is correct
Vitamin $B_{12}$ is absorbed in the ileum.

5.024  Decreased serum concentration of vitamin A

B. is correct
Bile acids that are necessary for fat digestion are primarily absorbed in the ileum. Loss of bile acids results in steatorrhea and decreased absorption of fat soluble vitamins.

**Use the following to answer questions 25-27.**

    A.    inhibits gastric motility
    B.    stimulates duodenal motility
    C.    stimulates gastric motility
    D.    inhibits duodenal motility
    E.    none of the above

5.025 Luteal phase of menstrual cycle

D. is correct
Progesterone is associated with the luteal phase of the menstrual cycle and inhibits gastrointestinal motility.

5.026 100 mN HCl in duodenum

A. is correct
Acid chyme in the small intestine stimulates the enterogastric reflex to inhibit gastric motility.

5.027 310 mM glucose in duodenum

A. is correct
Hypertonic and high caloric solutions in the duodenum inhibit gastric motility perhaps via the enterogastric reflex.

**Use the following to answer questions 28-30.**

    A.    colon
    B.    ileum
    C.    duodenum
    D.    jejunum
    E.    none of the above

5.028 Active absorption of bicarbonate

E. is correct
Bicarbonate is actively exchanged for chloride in the ileum and colon; there is no active absorption of bicarbonate in the gut.

5.029 Active absorption of bile salts

B. is correct
Active absorption of bile salts occurs in the ileum.

5.030 Primary site of iron absorption

C. is correct
The duodenum is the primary site of iron absorption.

**Use the following to answer question 31-32**

Your patient, a 42-year-old, white male develops an ulcer, nine (9) months following a vagotomy and partial gastrectomy for a bleeding duodenal ulcer. In addition to the pain of the ulcer, he also has profuse diarrhea. Three (3) years previously, he had a parathyroid tumor removed. Both basal and stimulated acid output is elevated.

5.031 The most likely diagnosis is:
    A. Gastrin
    B. VIP
    C. Gastrinoma
    D. VIPoma
    E. Hyperparathyroidism

C. is correct
Zollinger-Ellison syndrome or gastrinoma is diagnosed by hypersecretion of gastric acid resulting in duodenal ulceration and impairment of normal digestion and absorption through acidification of the duodenum resulting in diarrhea.

5.032 The hormone responsible for the changes noted in this case is:

A. is correct
This condition is due to gastrin producing tumor most often located in the pancreas.

**Use the following to answer questions 33-35.**

    A. may be ameliorated with cholestyramine
    B. excellent response to a gluten free diet
    C. common in person infected with *Helicobacter pylori*
    D. good response to milk withdrawal
    E. steatorrhea decreased with administration of digestive enzymes

5.033 Celiac sprue

B. is correct
Celiac sprue of gluten enteropathy is an absorption defect relieved by complete elimination of gluten from the diet.

5.034 Chronic pancreatitis

E. is correct
Chronic pancreatitis can lead to destruction of pancreatic acinar cells and hence digestive enzyme deficiency.

5.035 Duodenal ulcer

C. is correct
Presence of duodenal ulcers is strongly associated with *Helicobacter pylori* infection.

5.036 The presence of which of the following ions is necessary for active transport of sugars:
    A. potassium
    B. magnesium
    C. sodium
    D. mucosal cell
    E. adipose tissue

C. is correct
Transport of some sugars is facilitated by high concentration of sodium in lumen and vice versa. The transport of sodium from lumen to provide all the energy necessary to transport the hexose into the cell.

5.037 Following the intraluminal hydrolysis of fat, resynthesis of tryglycerides takes place in the:
    A. liver
    B. micelle
    C. microvillus
    D. mucosal cell
    E. adipose tissue

D. is correct
Resynthesis of triglycerides from free fatty acids and monoglyceride takes place in the mucosal cell after intramural hydrolysis.

5.038 The desire to defecate is initiated by:
    A. contraction of the external anal sphincter
    B. contraction of the internal anal sphincter
    C. distension of the sigmoid
    D. distension of the rectum
    E. contraction of the rectum

D. is correct
Distension of the rectum by fecal matter is the primary stimulus for defecation.

5.039 Your patient complained of constipation dating from birth, abdominal distension and occasional severe vomiting. Based on your knowledge of GI physiology, you would suspect:
   A. achalasia
   B. non-tropical sprue
   C. celiac disease
   D. megacolon
   E. rectal prolapse

D. is correct
Aganglionic megacolon is characterized by constipation, abnormal distension and anorexia.

5.040 The anatomical abnormality for this clinical picture is:
   A. hypertrophy of the descending colon
   B. colonic paralysis
   C. aganglionosis involving the distal colon
   D. rectal atresia
   E. absence of the internal rectal sphincter

C. is correct
Aganglionic megacolon (Hirschsprung's Disease) is due to a congenital absence of ganglion cells in the intramural plexuses in the distal colon.

**Use the following to answer question 41.**

A 25-year-old female patient complains of abdominal distension and diarrhea each time she drinks two glasses of milk, although she was able to drink milk as a child.

5.041 She most likely has:
   A. Zollinger-Ellison syndrome
   B. intestinal lactase deficiency
   C. allergy to milk
   D. hyperparathyroidism
   E. milk-alkali syndrome

B. is correct
Low lactase levels are associated with intolerance to milk (lactose intolerance).

5.042 The most likely prognosis of the problem described above would be:
   A. almost always fatal
   B. good if she survives the first 24 hours
   C. dependent on the outcome of surgery
   D. good responses to milk withdrawal
   E. bad unless exchange transfusion is performed

D. is correct
The problem of milk intolerance can be relieved by milk avoidance or by administration of commercial lactase preparations which alleviate the lactase deficiency.

**Use the following to answer question 43.**

A 38-year-old male school teacher presents with a six month history of excessive burping and nocturnal heartburn. There is no history of weight loss and physical exam is negative. A complete GI work-up with upper GI series and esophogram is negative except for a distended stomach.

5.043 The most likely cause of his problem is:
   A. gallbladder disease
   B. duodenal ulcer
   C. reflux esophagitis
   D. achalasia
   E. diffuse esophageal spasm

C. is correct
Regulation of acid into the mucosa of the lower esophagus may cause esophageal spasm resulting in pain and inflammation known as heartburn.

5.044  Secondary bile acids are:
   A.  activated by enterokinase
   B.  more soluble than primary bile acids
   C.  secreted only in adulthood
   D.  produced from primary bile acids by bacterial
       action in the intestine
   E.  not subject to enterohepatic circulation

D. is correct
Secondary bile acids are formed from primary bile acids by bacterial dehydroxylation and are actively reabsorbed from the ileum along with primary bile salts.

5.045  Obstructive jaundice without any liver damage would produce which combination of the following changes in bilirubin levels?

| Bilirubin in plasma | | Bilirubin in urine |
|---|---|---|
| Unconjugated | Conjugated | Conjugated |
| A.  increase | no change | increase |
| B.  no change | increase | increase |
| C.  decrease | decrease | decrease |
| D.  increase | increase | increase |
| E.  decrease | no change | increase |

B. is correct
Bilirubin, a bile pigment, is conjugated in the liver and is secreted in bile. Obstruction of the bile duct will result in no change in unconjugated bilirubin in the plasma but will cause a back-up of conjugated bilirubin into the plasma and subsequent appearance in the urine.

5.046  After secretion into the duodenum, the enzyme trypsinogen is converted into its active form, trypsin by:
   A.  enterokinase
   B.  procarboxypeptidase
   C.  lysolecithin
   D.  an alkaline pH
   E.  an acid pH

A. is correct
Proteolytic enzymes are secreted in pancreatic juice as inactive precursors. Trypsinogen is activated in the lumen by enterokinase.

5.047  Which of the following statements most accurately describes amino acid uptake from the intestinal lumen?
   A.  crosses the mucosal membrane by a process of
       passive diffusion
   B.  crosses the mucosal membrance by a carrier-
       mediated transport process requiring the
       concomitant transfer of Na+ in the same direction
   C.  crosses serosal membrane by a carrier-mediated
       transport process requiring the concomitant
       transfer of Na+ in the opposite direction
   D.  is actively transported across the serosal
       membrane against a concentration gradient
   E.  uptake is independent of high energy phosphate
       bonds in cell

B. is correct
The amino acid transport system requires the presence of sodium in the lumen and the mechanism of transport may be similar to that of glucose.

5.048  If excessive loss of gastric juice occurs due to chronic vomiting, the following are seen:
   A.  dehydration
   B.  hyperchloremia
   C.  alkalosis
   D.  all of the above
   E.  only A and C

E. is correct
Chronic vomiting can lead to dehydration hypochloremia, hypokalemia, and alkalosis.

5.049 The colon performs the following EXCEPT:
- A. $H_2O$ absorption is active
- B. actively absorbs $Na^+$ and secretes $K^+$
- C. exhibits mass movements of contents
- D. absorption is couple with $HCO_3$ secretion
- E. none of the above

A. is correct
Water absorption is passive throughout the gut.

5.050 Bile contains the following EXCEPT:
- A. lecithin
- B. cholesterol
- C. chymotrypsinogen
- D. electrolytes
- E. compounds in micelle complex

C. is correct
Chymotrypsinogen is a proteolytic proenzyme released by the pancreas and is not contained in bile.

5.051 After removal of the ileum, one observes a decrease in the following EXCEPT:
- A. in size of bile acid pool
- B. in fat content of feces
- C. in absorption of vitamin $B_{12}$
- D. all of the above
- E. only A and B

B. is correct
Resection of the ileum will result in a decrease in absorption of bile acids and vitamin $B_{12}$. Therefore, fecal fat content will increase.

5.052 Micelle formation is necessary for absorption and transport to intestinal epithelium of the following EXCEPT:
- A. cholesterol
- B. vitamin K
- C. vitamin D
- D. vitamin C
- E. vitamin E

D. is correct
Micelle formation is necessary for fat absorption including fat soluble vitamins A, D, E and K.

5.053 Functions that do not need extrinsic innervation are the following EXCEPT:
- A. initiation of swallowing
- B. stomach emptying
- C. pancreatic juice electrolyte secretion
- D. mixing and propulsion of small intestine

A. is correct
Muscles of the pharynx, upper esophageal sphincter and upper 1/3 of the esophagus are striated muscles innervated by motor neurons under voluntary control.

5.054 Failure to absorb vitamin $B_{12}$ from the gastrointestinal tract may result in pernicious anemia. The mucopolysaccharide or mucopolypeptide in the normal stomach mucus which combines with vitamin $B_{12}$ and makes it available for absorption by the gut is called:
- A. secretin
- B. intrinsic factor
- C. pancreozymin
- D. antihemophilic factor A
- E. pyridozine

B. is correct
Intrinsic factor is secreted by the oxyntic cell and combines with $B_{12}$ in the duodenum. When the I. F. and $B_{12}$ complex reaches the ileum the complex disassociates and $B_{12}$ is absorbed.

5.055  There are several important naturally occurring amino acids which do not occur in proteins. The amino acid which combines with bile acids in man is:
   A.  citruline
   B.  5-hydroxtryptophan
   C.  monoiodotryrosine
   D.  homocysteine
   E.  taurine

E. is correct
Bile acids are conjugated with either glycine or taurine and since the body contains more glycine than taurine, there is more glycocholate than taurocholate.

5.056  Which of the following peptides is(are) correctly paired with a hormone action?
   A.  secretin-pancreatic fluid and electrolyte secretion
   B.  motilin-gastric and duodenal migrating motor complexes (MMCs)
   C.  somatostatin-inhibition of gastrin release
   D.  cholecystokinin-pancreatic enzyme secretion
   E.  all of the above

E. is correct
All are correctly paired.

5.057  It is known that gastrin:
   A.  has actions similar to pepsin
   B.  is not secreted in its active form
   C.  reaches the stomach target cells via the circulation
   D.  initiates the conversion of pepsinogen to pepsin
   E.  none of the above

C. is correct
Gastrin is a small regulatory peptide hormone that is released into the circulation from G cells in the gastric antrum into the circulation where it travels to oxyntic cells to stimulate gastric acid secretion. Secondarily, it promotes pepsinogen secretion by gastric peptic (chief) cells.

5.058  It is known that secretin:
   A.  is synthesized by the pancreas
   B.  is synthesized by the small intestinal mucosa
   C.  directly neutralizes acid chyme from the stomach
   D.  has an optional pH of 8. 5
   E.  none of the above

B. is correct
Secretin is a regulatory peptide that is released by the intestinal mucosa in response to acid chyme to stimulate watery bicarbonate secretion from the exocrine pancreas.

5.059  The major factor that stimulates the release of GIP into the blood stream is:
   A.  parasympathetic stimuli
   B.  intraduodenal glucose
   C.  a stomach full of digested contents
   D.  peptones in the gastric chyme that enter duodenum
   E.  none of the above

B. is correct
GIP, glucose insulintropic-peptide, is released from intestinal mucosa in response to intraduodenal glucose.

5.060  Which of the following regarding the vagal innervation of the gut is(are) correct?
   A.  vagal innervation does not influence the enterogastric reflex
   B.  vagotomy eliminates the peristaltic contractions of the smooth muscle portion of the esophagus
   C.  the influence of the vagus is usually mediated via secondary neurons in the enteric nervous system
   D.  the migrating motor complex (MMC) is entirely dependent on vagal innervation
   E.  all of the above

C. is correct
Vagal parasympathetic innervation to the gut modulates intrinsic activity generated by the enteric nervous system within the walls of the GI tract. Extrinsic innervation to the tract can be cut and normal motility will remain due to the presence of this intrinsic nervous system.

5.061 Which of the following is (are) correct concerning intestinal motility?
   A. slow waves originate in the longitudinal muscle
   B. slow wave frequency increases in more distal portions of the small bowel
   C. feeding delays the migrating motor complex (MMC)
   D. segmenting contractions serve primarily to move food from the proximal to distal portions of the small bowel
   E. none of the above

E. is correct
Slow waves originate in the longitudinal muscles and their frequency decreases from proximal to distal portions of the small bowel. Segmentation contractions serve to mix food and expose chyme to absorbing cells but result in no net forward movement. Feeding stops the MMC.

5.062 Which of the following is (are) correct?
   A. 100-200 ml of saliva is secreted daily in most healthy people
   B. the composition of electrolytes in saliva is unchanged with different rates of secretion
   C. salivary secretion is primarily controlled by vasoactive intestinal peptide
   D. cutting the extrinsic innervation to the parotid and submandibular glands significantly reduces total salivary secretion
   E. none of the above

D. is correct
Salivary secretion is entirely under control of autonomic nerves. A total 1.0-1.5 liters of saliva are secreted daily and the electrolyte composition of saliva is altered during increased rates of secretion.

5.063 A likely reason why a "to and from" motion such as that encountered in an automobile ride over a bumpy road tends to cause nausea and vomiting is:
   A. too much air is swallowed and the stomach "bloats"
   B. the stomach is more likely to develop reverse peristalsis
   C. the cerebral cortex can no longer inhibit a tendency of the brainstem to cause vomiting
   D. vestibular reflexes excite a chemoreceptor trigger zone in the medulla
   E. none of the above

D. is correct
Motion sickness is associated with activation of the vestibular system and leads to neural activation of the medullary vomiting center via the chemoreceptor trigger zone.

5.064 A pyloric obstruction in an infant patient, which results in a prolonged bout of severe vomiting is likely to cause:
   A. metabolic acidosis
   B. metabolic alkalosis
   C. respiratory acidosis
   D. respiratory alkalosis
   E. none of the above

B. is correct
Metabolic alkalosis develops because of a loss of hydrogen ions in the vomitus.

5.065 The secretion of water and bicarbonate by the pancreas is primarily increased by:
   A. secretin
   B. villikinin
   C. cholecystokinin
   D. chymodenin
   E. motilin

A. is correct
Secretin stimulates ductal cells to secrete a watery bicarbonate solution.

5.066 Which of the following is correct concerning gastric acid secretion?
   A. histamine-1 receptor antagonists will block about 50% of acid secretion
   B. histamine-2 receptor antagonists will block 80-90% of acid secretion after a meal
   C. cimetidine will block histamine stimulated acid secretion but not affect gastrin stimulated acid secretion
   D. cimetidine will block histamine stimulated acid secretion but not affect vagal stimulated acid secretion
   E. none of the above

B. is correct
In the stomach histamine potentiates the effect of gastrin and acetylcholine on the parietal cell. Blockade of this potentiation results in an approximate 80% reduction of gastric acid secretion.

5.067 Which of the following is least important in the pathogenesis of reflux esophagitis?
   A. incompetence of the lower esophageal sphincter
   B. poor clearing of refluxed material from the lower esophagus
   C. the presence of acid and pepsin in the lower esophagus
   D. chronic gastroesophageal reflux
   E. swallowing difficulties associated with poorly chewed food

E. is correct
The lower esophageal sphincter is the major barrier to reflux. Effective clearing by secondary peristalsis is also important. Swallowing difficulties associated with poorly chewed food are not related to prolonged relaxation of the lower esophageal sphincter.

5.068 All of the following concerning the mucosa of the small intestine are correct EXCEPT:
   A. regeneration rate is very slow
   B. secretes enterokinase which has another protein enzyme as its substrate
   C. absorbs more than one liter of isotonic fluid per day
   D. is stimulated to secrete at a faster rate by parasympathetic activity
   E. folds, villi and microvilli of the epithelial surface yield an area estimated to be about 200 square meters

A. is correct
The regeneration rate of small intestinal mucosa is fast since the epithelium is a rapidly proliferating tissue.

5.069 The principal stimulus which causes the gallbladder to empty and the bile to be released is:
   A. cholecystokinin
   B. secretin
   C. enterogastrone
   D. enterokinin
   E. pepsin

A. is correct
Cholecystokinin is the principal humoral factor controlling gallbladder contraction and relaxation of the sphincter of oddi.

5.070 Which of the following statements is TRUE about the autonomic innervation of the gastrointestinal tube?
   A. Sympathetic innervation of the esophagus, stomach, small intestine, and large intestine is via branches of the vagus nerve
   B. Parasympathetic stimulation of the myenteric plexus causes a general increase in motility of the gut tube
   C. Acetylcholine is the neurotransmitter released at the nerve terminal of most postganglionic sympathetic fibers
   D. Sympathetic stimulation usually increases contraction of smooth muscle in the wall of the stomach and intestine
   E. Sympathetic innervation excites intrinsic neurons in the myenteric plexus

B. is correct
Parasympathetic stimulation to the gut is generally cholinergic and excitatory. A major area where parasympathetic neurons innervating the lower esophageal sphincter area release non-cholinergic inhibitory substances such as nitric oxide (NO) and vasoactive intestinal peptide (VIP).

5.071 All of the following are true with regard to swallowing EXCEPT:
   A. Lumenal pressure in the middle third of the esophagus is maintained higher than that in the stomach.
   B. The reflex act of deglutition is triggered when food comes in contact with the wall of the oropharynx
   C. Swallowing centers in the medulla and sends signals, via the 5th, 9th, 10th and 12th cranial nerves, to muscle cells in the pharynx and upper esophagus
   D. Initiation of deglutition, involving movement of food by the tongue, is under voluntary control, but the pharyngeal and esophageal phases of swallowing are reflex acts.
   E. Respiration is so vital that signals from the respiratory center in the medulla will always override those from the swallowing center

A. is correct
Lumenal pressure is maintained at a higher level than that in the stomach in the lower esophageal sphincter area not in the middle third of the esophagus.

5.072 The force required for evacuation of gastric contents during vomiting is generated by:
   A. contraction of the proximal stomach
   B. reverse peristalsis in the esophagus
   C. contraction of abdominal muscles
   D. reverse peristalsis in the stomach
   E. expiring against a closed glottis

C. is correct
Contraction of abdominal muscles act against a relaxed stomach to force gastric contents up into esophagus.

5.073 Contractions associated with the interdigestive myoelectric complex:
   A. are important in preventing bacterial overgrowth in the small intestine
   B. occur during the post-prandial period
   C. are regulated by secretin
   D. are segmentation type contractions
   E. occur only in the stomach

A. is correct
The interdigestive myoelectric complex or the migrating motor complex perform a housekeeping function by sweeping acid chyme from the stomach to the ileum.

5.074 Diarrhea is caused by the following EXCEPT:
   A. reduced digestion and absorption of fatty food materials
   B. elevated absorption of water and ions in the large intestine
   C. damage to absorptive cells in the small intestine by cholera toxin
   D. bacterial infection of the large intestine causing enteritis
   E. extreme parasympathetic stimulation (such as from emotional disturbance) to intestine

B. is correct
Diarrhea is caused by overwhelming the absorptive capacity of the colon resulting in a watery stool.

5.075 *Helicobacter pylori:*
   A. is most often found in the descending colon
   B. accounts for many ulcers that respond to antibiotic treatment
   C. refers to colonies of *Escherichia coli* that are found in spiral glands in the pyloric stomach
   D. is rarely found in adult humans
   E. usually cannot grow in the acidic environment of the stomach

B. is correct
*Helicobacter pylori* infestation is associated with gastrointestinal ulcer formation that responds to antibiotic therapy.

5.076 All of the following statements concerning Zollinger-Ellison syndrome are true EXCEPT:
   A. it results in gastric ulcers
   B. tumors in pancreatic islets are commonly found
   C. circulating gastrin levels are elevated
   D. mucus production by goblet cells is depressed
   E. gastric glands secrete abnormally high levels of HCl

D. is correct
Zollinger-Ellison or gastrinoma is produced by a gastrin releasing tumor usually in the pancreas resulting in excessive gastric acid secretion and ulcer formation.

5.077 All of the following are absorbed by $Na^+$-dependent secondary active transport EXCEPT:
   A. D-glucose
   B. D-galactose
   C. D-fructose
   D. L-alanine
   E. L-phenylalanine

C. is correct
Fructose is transported by facilitated diffusion not associated with the presence of sodium.

5.078 Which receptor on the parietal cell is blocked by cimetidine?
   A. somatostatin
   B. gastrin
   C. acetylcholine
   D. motilin
   E. histamine

E. is correct
Cimetidine (tagamet) is a specific histamine-2 receptor blocker.

5.079 All of the following are true concerning lipid digestion EXCEPT:
   A. little oral digestion occurs
   B. extensive gastric digestion occurs
   C. extensive small intestinal digestion occurs
   D. lipids are taken up by diffusion
   E. bile salts are required to emulsify dietary lipids

B. is correct
Lingual lipase in saliva begins fat digestion in the oral cavity but extensive lipid digestion does not occur in the gastric lumen.

5.080  Your patient presents with a history of constipation dating from infancy. An abdominal film confirms that he is impacted with stool and a barium enema is performed and demonstrates a dilated colon with contraction of the distal sigmoid colon and rectum. Which of the following is correct?
  A. this condition is caused by the congenital absence of sympathetic innervation of the esophagus
  B. there is an absence of relaxation of sphincter of the pyloric
  C. a full-thickness biopsy of the rectal mucosa will demonstrate an absence of intramural ganglia
  D. the dilated segments of colon are aganglionic
  E. the treatment of choice is a high fiber diet supplemented with laxatives

C. is correct
The condition is congenital megacolon or Hirschsprung's Disease which is characterized by a permanently constricted segment with dilation of the bowel above the constriction.

5.081  All of the following are ways in which the pancreas protects itself against the potentially harmful effects of its own digestive enzymes EXCEPT:
  A. proteolytic enzymes are segregated from the cytoplasm of the acinar cell by membranes
  B. those enzymes which can digest membranes are synthesized and secreted as inactive zymogens
  C. the activating enzyme is located away from the pancreas
  D. proteolytic enzymes are the smallest component of pancreatic secretory proteins
  E. the acinar cell synthesizes a trypsin inhibitor which blocks any trypsin inadvertently present in the acinar cell

D. is correct
Proteolytic enzymes make up 80% of the total pancreatic protein secretion.

5.082  Which of the following is correct regarding achalasia?
  A. the pressure in the lower esophageal sphincter (LES) is lower than the stomach
  B. abnormalities in LES function are associated with degeneration of excitatory neurons
  C. the LES is maintained in a contracted state by circulating progesterone
  D. the motility in the distal 2/3 of the esophagus is characterized by weak, simultaneous (nonperistaltic) contractions following swallows
  E. the narrowing of the distal esophagus is due to reflux esophagitis with peptic stricture

D. is correct
Achalasia is characterized by failure of the LES to relax along with aperistalis (no esophageal peristaltic wave).

5.083  Chronic vomiting often results in all of the following EXCEPT:
  A. metabolic acidosis
  B. metabolic alkalosis
  C. hypokalemia
  D. hypochloremia
  E. dehydration

A. is correct
The loss of gastric acid through prolonged vomiting results in metabolic alkalosis characterized by increased plasma bicarbonate and elevated plasma pH.

5.084 A patient visits your office in Kenya complaining of lower left quadrant pain. A barium enema demonstrates numerous diverticula. Which of the following would you expect to be correct?
   A. the patient is a life-long resident of Africa who has just moved to Kenya from Zambia
   B. the patient is a farmer from the Kenyan country-side
   C. the patient regularly eats a diet high in fiber
   D. the patient regularly eats a diet deficient in fiber and is a recent immigrant from Great Britian
   E. the patient's symptoms are psychogenic in origin

D. is correct
Diverticulitis is almost unknown in native Africans who generally consume high fiber diets.

5.085 A low intragastric pH (<2. 0):
   A. inhibits the release of gastrin
   B. inhibits the release of histamine
   C. inhibits peptic activity
   D. promotes gastric motility
   E. reduces irritation by aspirin

A. is correct
Low gastric pH inhibits further release of gastrin probably via somatostatin.

5.086 The pathway from the intestinal lumen to the circulating blood for a short-chain fatty acid less than 10 carbons in length is:
   A. intestinal mucosal cell -> chylomicrons -> lymphatic duct -> systemic venous blood
   B. intestinal mucosal cell -> hepatic portal vein -> systemic venous blood
   C. aqueous channel -> chylomicrons -> portal vein -> systemic venous blood
   D. aqueous channel -> chylomicrons -> lymphatic duct -> systemic venous blood
   E. lumenal digestion -> intestinal mucosal cell -> LDL -> hepatic portal vein -> systemic venous blood

B. is correct
Short-chain fatty acids do not have to be digested and directly enter the portal circulation.

5.087 All of the following are correct concerning splanchnic circulation EXCEPT:
   A. the splanchnic organs make up less than 10% of body mass but receive 25% of the resting cardiac output
   B. splanchnic organs are unable to autoregulate in the face of falling blood pressures
   C. secretin produces a decrease in mucosal blood flow
   D. under conditions of severe ischemia the counter-current exchange of oxygen in the intestinal villus is exaggerated
   E. despite continuous infusion of vasoconstrictor drugs in the superior mesenteric artery blood flow will return to preconstrictor values

B. is correct
Although some splanchnic organs cannot autoregulate, the small intestine demonstrates some autoregulatory ability to maintain constant flow in the face of falling pressures.

5.088 All of the following are correct concerning salivary secretion EXCEPT:
  A. the primary acinar cell secretion is isotonic with the plasma
  B. the duct epithelium actively absorbs sodium
  C. the ducts are quite permeable to water and water flows in by solvent drag
  D. chloride ions move out of the duct lumen either in exchange for bicarbonate or by passive diffusion
  E. the osmolality and electrolyte composition of saliva increases at high rates of secretion

C. is correct
Salivary ducts are impermeable to water, thereby making salivary secretion the only hypotonic digestive secretion.

5.089 In 1902, Bayliss and Starling discovered the first blood-borne chemical messenger which they called a hormone. One of its major actions is to:
  A. stimulate gastric acid secretion
  B. inhibit the contraction of the lower esophageal sphincter
  C. stimulate pancreatic bicarbonate secretion
  D. inhibit pepsinogen release
  E. stimulate gallbladder contraction

C. is correct
In 1902, Bayliss and Starling discovered the first hormone, secretin, that stimulates a watery bicarbonate secretion from the pancreas.

5.090 Which of the following is correct concerning gastric secretion?
  A. vagal stimulation increases the amount of somatostatin released into the gastric circulation
  B. the intestinal phase is responsible for the largest percentage of secretion during a meal
  C. in order for gastrin to be released, the pH of the gastric contents must be raised above 3.0
  D. acid peptic digestion of protein in the stomach releases peptides and amino acids, both of which stimulate bombesin release
  E. the largest amount of gastrin released during a meal is from the intestinal G cells

C. is correct
Gastrin release is inhibited at gastric pH levels of 2.0 and below.

**Use the following to answer questions 91-95.**

  A. bombesin                               F. motilin
  B. chymodenin                             G. secretin
  C. cholecystokinin                        H. somatostatin
  D. gastrin                                I. substance P
  E. gastric inhibitory peptide (GIP)       J. vasoactive intestinal peptide (VIP)

5.091 Peptide responsible for contraction of gallbladder secretion

C. is correct
Cholecystokinin is the peptide primarily responsible for gallbladder contraction.

5.092 Secreted by G-cells of the antrum

D. is correct
Gastrin is secreted by the G-cells of the antrum.

5.093  Believed responsible for relaxation of lower esophageal sphincter

J. is correct
VIP released by inhibitory neurons is believed to be responsible for lower esophageal relaxation.

5.094  Stimulates a watery bicarbonate secretion of the pancreas

G. is correct
Secretin stimulates the watery bicarbonate portion of pancreatic secretion.

5.095  Inhibits pancreatic enzyme secretion

H. is correct
Somatostatin, the universal inhibitory peptide, inhibits pancreatic enzyme secretion.

**Use the following to answer questions 96-100.**

A. cyanocobalamin      E. intrinsic factor
B. chymotrypsin        F. pepsin
C. ptyalin             G. trypsin
D. enterokinase        H. taurocholate

5.096  Secreted by parietal cells of the stomach

E. is correct
Intrinsic factor is secreted by the acid producing cells; the parietal or oxytic cells of the stomach.

5.097  Begins the breakdown of starch in the mouth

C. is correct
Ptyalin attacks the l, 4 glucosidic linkages of starch in the mouth.

5.098  Activates trypsinogen

D. is correct
Enterokinase from the small intestinal mucosa initally activates trypsinogen to trypsin.

5.099  Gastric acid secretion stimulates activity

F. is correct
Pepsinogen is converted to its active form pepsin by the acidic environment of the stomach.

5.100  Undergoes enterohepatic circulation

H. is correct
Taurocholate is a primary bile acid that returns to the liver by enterohepatic circulation to be resecreted to assist in the digestion of a fatty meal.

# SECTION 6:    ENDOCRINE

6.001  All of the following pairs of items are correctly associated EXCEPT:
  A. Leydig cell - androgen secretion
  B. follicle stimulating hormone - Sertoli cell
  C. Sertoli cell - blood testis barrier
  D. spermatogonia - acrosome formation
  E. androgen binding protein - Leydig cell

E. is correct
The Sertoli cells secrete androgen binding protein, inhibin and Mullerian inhibitory substance

6.002  Testes normally enter the scrotal sac during the:
  A. 2nd month of development
  B. 4th month of development
  C. 6th month of development
  D. 9th month of development
  E. 1st postnatal month

D. is correct
Testicular descent is normally complete by the 9th month just prior to parturition

6.003  Testicular androgens are produced in:
  A. Sertoli cells
  B. Leydig cells
  C. spermatocytes
  D. tubular cells
  E. none of the above

B. is correct
Testosterone, the primary testicular androgen, is produced in the intestinal cells of Leydig

6.004  Normal sexual development in the human male fetus depends on:
  A. the presence of Mullerian inhibitory substance (MIS)
  B. secretion of testosterone
  C. testosterone receptors
  D. presence of five (5) alpha reductase
  E. all of the above

E. is correct
Male internal and external genitalia develop when functional testes are present to secrete testosterone, MIS and when there is adequate $5\ \alpha$ reductase and no androgen resistance present

6.005  The primary function of seminiferous tubules is to produce:
  A. immotile sperm
  B. androgens
  C. progesterone
  D. estrogens
  E. motile sperm

A. is correct
Spermatozoa achieve motility during their further maturation in the epididymis

6.006  A patient who has been taking testosterone for several years may have:
  A. large muscles
  B. low sperm production rate
  C. low plasma gonadotropin levels
  D. small testes
  E. all of the above

E. is correct
Testosterone will produce anabolic effects such as muscle growth but also androgenic effects such as negative feedback on the hypothalamic - pituitary system to inhibit gonadotropin production

6.007 Which of the following is/are function(s) of
Sertoli cells?
    A. secretion of androgen-binding protein (ABP)
    B. nourishment of germ cells through spermatogenesis
    C. probable source of inhibin
    D. formation of the blood testis barrier
    E. all of the above

E. is correct
Sertoli cells secrete androgen binding protein,
inhibin, for the blood-testis barrier by their
cellular junctions, and allow spermatozoa
formation through maturation in their folds.

6.008 Your patient presents clinically with a cervical
mucus sample - it is thin, alkaline and dries in a fern-like
pattern. From this you can infer that your patient:
    A. likes botany
    B. is pregnant
    C. is about to menstruate
    D. is near ovulation
    E. all of the above

D. is correct
Estradiol causes increased production of a watery
mucus by the cervix which dries in a fern-like
pattern. Elevated estrogen levels at midcycle are
believed to trigger the LH surge and follicle
rupture.

6.009 In the secretory phase of the uterine cycle there is:
    A. an absence of a positive feedback effect of
       estrogen to induce an LH peak
    B. a scant, thick mucus produced by cervical gland
    C. an increased secretion of progesterone and
       estrogen by the corpus luteum
    D. a secretion of endometrial glands known as
       uterine "milk"
    E. all of the above

E. is correct
The secretory phase of the uterine cycle is
characterized by increased plasma levels of
estrogen and progesterone, increased endometrial
gland secretion and thick cervical mucus
stimulated by progesterone and inhibition of the
estrogen positive feedback by progesterone.

6.010 Estrogen does:
    A. increase development of the primary ducts in the
       mammary gland
    B. decrease responsiveness of uterine muscle to
       spontaneous contractions
    C. increase circulating LH and FSH when given
       after ovariectomy
    D. decrease growth of the uterine endometrium
    E. all of the above

A. is correct
Estrogen increases ductal development of
mammary glands, sensitizes the uterus to
spontaneously contract and stimulates
endometrial growth.

6.011 Growth of axillary and pubic hair in normal
females is primarily due to:
    A. adrenal androgens
    B. adrenal estrogens
    C. ovarian progesterone
    D. ovarian estrogens
    E. ovarian androgens

A. is correct
Development of pubic and axillary hair in the
female is primarily due to androgen rather than
estrogen. Most androgens in the female come
from the adrenal cortex.

6.012 The gradual decrease in number of ova with age
in human ovaries is due to ovulation and:
    A. follicular atresia and a gradual decrease in
       production of new oocytes after birth
    B. an increase in the rate of follicular atresia with a
       constant production of new oocytes after birth
    C. follicular atresia without any production of new
       oocytes after birth
    D. increase in the rate of follicular atresia with a
       decrease in production of new oocytes after birth
    E. the action of prolactin

C. is correct
No new ova are formed after birth and
continuing atresia reduces the number of ova
during development and adulthood.

6.013  In the human, fertilization of the ovum usually occurs:
   A. during menstruation
   B. after the ovum has traveled to the uterus
   C. in the oviducts
   D. 4-6 days after ejaculation
   E. while the ovum is still contained within the follicle

C. is correct
Fertilization of the ovum by sperm usually occurs the mid-portion of the oviduct in humans.

6.014  Measurement of gonadotropins in serum can give an indication of the function of all of the following EXCEPT:
   A. anterior pituitary
   B. uterus
   C. ovary
   D. hypothalamus
   E. testis

B. is correct
Gonadotropin secretion is dependent upon function of the hypothalamus and pituitary. Gonadotropins stimulate gonadal function and do not directly affect sex accessory glands.

6.015  All of the following tests would be helpful for workup of amenorrheic women who have normal physical examinations EXCEPT:
   A. serum FSH, LH
   B. biopsy of endometrium
   C. urinary 17-ketosteroids
   D. response to exogenous progesterone
   E. all of the above

C. is correct
Most urinary 17-ketosteroids are weak androgens originating either from the adrenal gland.

6.016  The menstrual cycle:
   A. begins after the first pregnancy and ends at menopause
   B. has an average length of nine (9) months
   C. results from interaction of hypothalamus, pituitary, and adrenal gland
   D. continues through pregnancy
   E. none of the above

E. is correct
The menstrual cycle begins with the first day of bleeding associated with endometrial sloughing. The cycle is approximately 28 days in length with bleeding occurring on days 1-5 with an approximate 30ml blood loss. The adrenal gland is not directly involved.

6.017  The amenorrhea after menopause is caused by the inability of the:
   A. anterior pituitary to produce gonadotropins
   B. hypothalmus to produce GnRH.
   C. endometrium to respond to estradiol
   D. endometrium to respond to progesterone
   E. ovaries to respond to gonadotropins

E. is correct
The ovaries become unresponsive to gonadotropins with advancing age that is associated with a significant decline in the number of primordial follicles

6.018  The control of milk ejection:
   A. includes the posterior lobe of the pituitary gland
   B. involves contraction of myoepithelial cells of mammary gland
   C. includes oxytocin in its efferent component
   D. requires intact sensory nerves from the mammary gland
   E. all of the above

E. is correct
Oxytocin causes contraction of the myoepithelial cells with consequent ejection of milk through the nipple. The release of oxytocin is a neuroendocrine reflex that requires suckling stimulus. Oxytocin can also be released as an emotional response to baby.

6.019 During the months preceding the menopause:
  A. FSH is elevated more than LH
  B. both FSH and LH are elevated
  C. FSH elevation precedes LH elevation
  D. all of the above
  E. A and C only

D. is correct
Due to a decline in the negative feedback of ovarian steroids, estrogen and progesterone, the secretion of FSH and LH is increased with FSH elevation preceding LH elevation

6.020 All of the following are contraindications to estrogen replacement therapy for the postmenopausal woman EXCEPT:
  A. venous thrombosis
  B. liver impairment
  C. hyperlipidemia
  D. depression
  E. smoking

D. is correct
Estrogen replacement therapy has been shown to enhance mood and well-being and libido.

6.021 All of the following hormones are stimulated by low serum glucose EXCEPT:
  A. epinephrine
  B. growth hormone
  C. cortisol
  D. thyroxine
  E. glucagon

D. is correct
Hypoglycemia has no effect on thyroxine production.

6.022 A hyperphosphatemic and hypocalcemic patient given an injection of parathyroid hormone that increases his urinary $PO_4$ excretion and decreases urinary $Ca^{++}$ excretion is probably suffering from:
  A. primary hypoparathyroidism
  B. pseudohypoparathyroidism
  C. pseudopseudohypoparathyroidism
  D. vitamin D deficiency
  E. Type I vitamin D dependence

A. is correct
PTH increases urinary phosphate excretion and calcium reabsorption. In pseudohypoparathyroidism, the tissues do not respond to PTH administration.

6.023 Which of the following hormones is used to test for pregnancy?
  A. 17β-estradiol
  B. progesterone
  C. human chorionic gonadotropin
  D. follicle stimulating hormone
  E. luteinizing hormone

C. is correct
Human chorionic gonadotropin is present in urine as early as a week after conception and is the basis of all laboratory tests for pregnancy.

6.024 All of the following will increase glucagon secretion EXCEPT:
  A. hypoglycemia
  B. increased protein intake
  C. infection
  D. an oral glucose tolerance test
  E. parasympathetic stimulation of pancreatic α cells

D. is correct
An increase in plasma glucose will inhibit glucagon secretion.

6.025 All the following will increase thyroid stimulating hormone (TSH) secretion EXCEPT:
  A. iodide deficiency
  B. an increased plasma concentration of $T_3$
  C. thyrotropin releasing hormone
  D. propylthiouracil
  E. thyroidectomy

B. is correct
Decreased plasma concentration of $T_3$ inhibits increased TSH stimulated by TRH.

6.026  Adrenocorticotropic hormone secretion is increased by all of the following EXCEPT:
   A. noxious stimuli
   B. electrical stimulation of the hypothalamus
   C. increased secretion of vasopressin
   D. aldosterone administration
   E. adrenatectomy

D. is correct
Aldosterone administration would not increase ACTH secretion. After hypophysectomy aldosterone secretion is normal.

6.027  A test of a patient's serum shows an estrogen spike. Which of the following should occur with the next three days?
   A. menses will begin
   B. ovulation
   C. the uterus will have actively proliferating glands
   D. menses will end
   E. the corpus luteum will involute

B. is correct
A peak in estrogen production for a minimum of 36 hours exerts a positive feedback on the pituitary to induce a gonadotropin surge that results in ovulation.

6.028  All of the following will lead to increased serum calcium EXCEPT:
   A. parathyroid hormone's action on renal tubules
   B. vitamin D's action on the gastrointestinal tract
   C. parathyroid hormone's action on the bone
   D. calcitonin
   E. 1-hydroxylation of 25-OH cholecalciferol

D. is correct
Calcitonin is a calcium lowering hormone produced by the parafollicular (c) cells of the thyroid gland.

6.029  Which of the following will increase growth hormone secretion:
   A. increased serum concentration of somatomedin C
   B. increased anterior pituitary concentration of somatostatin
   C. strenuous exercise
   D. complete sectioning of the pituitary stalk
   E. following infusion of glucose

C. is correct
Growth hormone secretion is stimulated by: 1) conditions such as hypoglycemia and fasting; 2) conditions that increase amino-acids in plasma; and 3) stressful stimuli. Somatostatin inhibits GH release and somatomedian C negatively feeds back to control GH secretion.

6.030  All of the following will increase secretion of aldosterone EXCEPT:
   A. increased serum concentration of angiotensin II
   B. hyperkalemia
   C. sympathetic renal nerve stimulation
   D. secretion of atrial natriuretic peptide
   E. renin secretion

D. is correct
Atrial natriuretic peptide inhibits renin secretion and decreases the responsiveness of the zona glomerulosa to angiotensin II.

6.031  Which of the following hormones promotes, either directly or indirectly, gluconeogenesis?
   A. follicle stimulating hormone
   B. insulin
   C. adrenocorticotropic hormone
   D. luteinizing hormone
   E. aldosterone

C. is correct
ACTH stimulates the release of adrenal glucocorticoids which stimulate increased gluconeogenesis.

**6.032** Which of the following is an important endocrine function of the kidney?
- A. 24 hydroxylation of 1-OH cholecalciferol
- B. erythropoietin production
- C. aldosterone conjugation to glucuronic acid
- D. a high concentration of angiotensin converting enzyme
- E. stimulation of $H^+$ retention by renin

**B. is correct**
Most erythropoietin in the body comes from the kidneys. Nephrectomy can therefore lead to anemia development because of the loss of erythropoietin, the major erythropoietic hormone.

**6.033** Insulin increases glucose uptake in:
- A. renal tubule cells
- B. neurons in the cerebral cortex
- C. muscle
- D. erythrocytes
- E. the mucosa of the small intestine

**C. is correct**
Although glucose enters all cells, insulin increases glucose uptake in muscle cells by increasing the number of glucose transporters in the cell membrane.

**6.034** The initial physiological effects of cortisol are due to:
- A. increased intracellular cAMP
- B. interaction with a plasma membrane receptor
- C. stimulation of $PIP_2$ metabolism
- D. binding to an intracellular receptor
- E. increased phosphoprotein phosphatase activity

**D. is correct**
The effects of glucocorticoids are initiated by binding to intracellular glucocorticoid receptors that promote DNA transcription leading to protein synthesis that alters cell function.

**6.035** The major hormone secreted by the ovarian follicle is:
- A. progesterone
- B. estriol
- C. 17β-estradiol
- D. estrone
- E. androstenedione

**C. is correct**
17β-estradiol is the major hormone of ovarian follicles. Androstenedione is aromatized to estrone in the periphery. Some estrone is metabolized to estriol primarily in the liver.

**6.036** All of the following are properties of thyroxine ($T_4$) EXCEPT:
- A. increases cellular $O_2$ function
- B. is composed of amino acid subunits
- C. binds to serum TBP with less affinity than $T_3$
- D. increases transcription of specific genes by binding to an intracellular receptor
- E. is considered to be a prohormone for $T_3$

**C. is correct**
Most of $T_4$ in the circulation is bound to TBP while $T_3$ exhibits lesser binding. The lower binding of $T_3$ to TBP correlates with the fact that $T_3$ has a shorter half-life than $T_4$.

**6.037** The primary regulator of testicular steroidogenesis is:
- A. prolactin
- B. follicle stimulating hormone
- C. testosterone
- D. luteinizing hormone
- E. dihydrotestosterone

**D. is correct**
Testosterone production by the testicular interstitial cells of Leydig is primarily regulated by luteinizing hormone.

**6.038** After glucagon binds to its receptor, which of the following events first occurs?
- A. protein kinase activity increases
- B. glycogen synthetase is phosphorylated
- C. phosphodiesterase activity decreases
- D. adenylate cyclase activity increases
- E. cAMP levels increase

**D. is correct**
When glucagon binds to its receptor on live cells it activates adenylate cyclase and thereby increases intracellular cyclic AMP.

6.039 Soon after insulin treatment ceases, a diabetic patient will exhibit:
A. a fall in blood glucose
B. increased plasma $HCO_3$
C. increased nitrogen excretion
D. absence of urinary ketone bodies
E. decreased urine volume

C. is correct
The diabetic patient has accelerated protein catabolism leading to negative nitrogen balance through increased nitrogen excretion. Insulin prevents this protein breakdown.

6.040 The intravenous administration of a calcium solution results in:
A. decreased secretion of calcitonin
B. increased urinary excretion of phosphate
C. decreased synthesis of Vitamin $D_3$
D. decreased formation of 1, 25-dihydroxy vitamin $D_3$

D. is correct
The formation of 1,25-dihydroxycholecalciferol is regulated in a feedback manner by plasma $Ca^{++}$ and $PO_4^{3-}$ levels.

6.041 All of the following are effects of growth hormone on body metabolism EXCEPT:
A. increasing the rate of protein synthesis
B. decreasing the rate of use of carbohydrate
C. decreasing the mobilization of fats
D. increasing the production of carbohydrate
E. increasing the use of fats for energy

C. is correct
Growth hormone increases lipolysis, protein synthesis and epiphyseal growth and decreases insulin sensitivity

6.042 All of the following increase insulin and glucagon secretion EXCEPT:
A. administration of carbachol
B. a rise in plasma glucose followed by a fall in plasma glucose
C. infusion of somatostatin
D. dietary protein intake
E. administration of β-adrenergic agonists

C. is correct
Somatostatin is a potent inhibitor of both insulin and glucagon.

6.043 An adrenal 11 β-hydroxylase deficiency will cause all of the following EXCEPT:
A. hyperplasia of the adrenal glands
B. increased secretion of ACTH
C. increased urinary 17-ketosteroid excretion
D. decreased urinary 17-OH corticosteroid excretion
E. decreased cortisol production

D. is correct
A 11 β-hydroxylase deficiency increases production of 17-OH corticosteroids leading to increased urinary excretion.

6.044 Somatostatin inhibits all the following EXCEPT:
A. vasopressin
B. thyroid stimulating hormone
C. insulin
D. growth hormone
E. gastrin

A. is correct
Somatostatin was first identified to inhibit growth hormone secretion but also inhibits, TSH, insulin and gastrin, but does not affect vasopressin release.

6.045 One factor to differentiate between primary and secondary adrenal insufficiency is:
A. hyponatremia
B. inability to excrete a water load
C. increased skin pigmentation
D. hyperkalemia
E. decreased glomerular filtration rate

C. is correct
Adrenal insufficiency caused by atrophy of the adrenal cortex elevates circulating ACTH. Patients with this problem develop hyperpigmentation due to the MSH activity of the ACTH, beta and gamma lipotropin in blood.

6.046 All of the following describe Cushing's disease EXCEPT:
- A. decreased collagen formation in skin and
- B. hypoglycemia
- C. decreased activity of phospholipase A$_2$
- D. increased supraclavicular fat accumulation
- E. decreased helper T-cell numbers

B. is correct
Excess plasma glucocorticoid levels lead to protein catabolism. Protein catabolism leads to increased amino acids in plasma which are converted into glucose in the liver and result in hyperglycemia.

6.047 All of the following conditions are associated with increased metabolic rate EXCEPT:
- A. elevation of plasma levels of T$_4$
- B. presence of peripheral vasodilation
- C. the development of fever
- D. increased muscular activity
- E. increased secretion of catecholamines

B. is correct
Increased metabolic rate is associated with heat production not heat loss that would be produced by peripheral vasodilation.

6.048 Obesity results from a lesion in which of the following hypothalamic nuclei?
- A. supraoptic
- B. ventromedial
- C. dorsomedial
- D. suprachiasmatic
- E. medial preoptic

B. is correct
Ventromedial hypothalamic lesions induce hyperphagia and subsequent obesity.

6.049 All of the following hormones are ketogenic EXCEPT:
- A. growth hormone
- B. cortisol
- C. epinephrine
- D. glucagon
- E. insulin

E. is correct
Insulin is anti-ketogenic and is a major lipogenic hormone.

6.050 All of the following hormones induce gluconeogenesis EXCEPT:
- A. growth hormone
- B. cortisol
- C. epinephrine
- D. glucagon
- E. insulin

E. is correct
Insulin is a major glycogenic hormone and does not induce gluconeogenesis.

6.051 The secretion of all of the following hormones is under control of hypothalamic releasing hormones EXCEPT:
- A. adrenocorticotropic hormone (ACTH)
- B. follicle stimulating hormone
- C. growth hormone
- D. luteinizing hormone
- E. prolactin

E. is correct
Prolactin secretion is tonically inhibited by the hypothalamus by a prolactin inhibitory hormone (PIH) thought to be dopamine.

6.052 Which of the following hormones will stimulate the release of TSH and prolactin?
- A. antidiuretic hormone (ADH)
- B. corticotropin releasing hormone (CRH)
- C. gonadotropin releasing hormone (GnRH)
- D. somatostatin (SRIF)
- E. thyrotropin releasing hormone (TRH)

E. is correct
TRH stimulates the secretion of both prolactin and TSH

6.053 The secretion of which of the following hormones is decreased by hypokalemia:
- A. aldosterone
- B. angiotensin II
- C. angiotensinogen
- D. renin
- E. vasopressin (ADH)

A. is correct
Aldosterone increases the retention of sodium and promotes the excretion of potassium. Thus, hypokalemia will result in decreased aldosterone release.

6.054 All of the following hormones increase throughout the duration of pregnancy EXCEPT:
- A. estriol
- B. human placental lactogen
- C. estrone
- D. human chorionic gonadotropin
- E. progesterone

D. is correct
Human chorionic gonadotropin (hCG) is produced by the syncytiotrophoblast soon after implantation and serves to bridge the gap between ovarian and placental maintenance of pregnancy. hCG reaches its highest levels in the first trimester of pregnancy then declines.

6.055 All of the following statements regarding vitamin D are true EXCEPT:
- A. multiple biologically active metabolites of vitamin D are known to exist
- B. vitamin D is converted 25-hydroxycholecalciferol by the liver
- C. 1,25-dihydroxycholecalciferol is believed to be the metabolically active form of vitamin D in the intestines
- D. 1,25-dihydroxycholecalciferol is made in the liver
- E. metabolites of vitamin D have selective biological activity such as effecting intestinal calcium transport.

D. is correct
1, 25-dihydroxycholecalciferol is synthesized in the kidney not the liver.

6.056 Which of the following are correct for blood sugar homeostasis?
- A. somatotropin always acts to lower blood sugar
- B. insulin and glucagon always react in opposite ways
- C. growth hormone increases insulin release and therefore lowers blood sugar
- D. glucagon raises blood sugar by decreasing cyclic AMP in the liver
- E. sympathetic nervous system stimulation inhibits both glucagon and insulin and results in lowered blood sugar

B. is correct
Insulin is glycogenic, antigluconeogenetic, antilipolytic and antiketotic, whereas glucagon is glycogenolytic, gluconeogenic, lipolytic and ketogenic. Thus, insulin and glucagon have opposite actions.

6.057 All of the following hormones either increase cyclic AMP concentration or stimulate the activity of the enzyme adenylate cyclase EXCEPT:
- A. adrenocorticotropic hormone (ACTH)
- B. luteinizing hormone (LH)
- C. follicle stimulating hormone (FSH)
- D. insulin
- E. melanocyte stimulating hormone (MSH)

D. is correct
Insulin secretion is probably produced by stimuli increasing intracellular $Ca^{++}$.

6.058 The mechanism of action of glucagon is due to:
A. increasing the absorption of glucose from the intestine
B. increasing the release of insulin from the pancreas
C. increasing the formation of cyclic AMP in the liver
D. inhibiting the uptake of glucose by tissues
E. decreasing insulin release

C. is correct
Glucagon exerts its action by increasing intracellular cyclic AMP after first activating adenylate cyclase.

6.059 Which of the following will increase the basal metabolic rate:
A. insulin
B. thyroxine
C. cortisone
D. ascorbic acid
E. glucagon

B. is correct
Thyroxine increases the oxygen consumption of almost all metabolically active tissues.

6.060 The signs and symptoms of Cushing's syndrome are primarily due to an excess secretion of:
A. testosterone
B. insulin
C. growth hormone
D. cortisol
E. calcitonin

D. is correct
The signs and symptoms are primarily due to excess secretion of glucocorticoids of which cortisol is the primarily example in humans.

6.061 Which of the following hormones act to increase circulating levels of free fatty acids?
A. progesterone
B. growth hormone
C. thyroid stimulating hormone
D. oxytocin
E. calcitonin

B. is correct
Growth hormone is ketogenic because it increases circulating free fatty acid levels providing a source of energy during fasting and stress.

6.062 Human chorionic gonadotropin (hCG) is an analogue of:
A. testosterone
B. luteinizing hormone
C. follicle stimulating hormone
D. progesterone
E. 17β-estradiol

B. is correct
hCG is primarily luteinizing and luteotropic having little FSH activity. It appears to act on the same receptor as LH.

6.063 With respect to human sexuality, androgens:
A. are precursors for pheromones
B. are required for libido in both sexes
C. play no role in sexual desire in either sex
D. provide feedback of luteinizing hormone in the female
E. regulate libido in males but not in females

B. is correct
Androgens are thought to be responsible for libido in both sexes.

6.064 The most common known cause of colloid goiter is:
A. Hashimoto's disease
B. carcinoma
C. iodine deficiency
D. adenoma
E. hyperthyroidism

C. is correct
Iodine deficiency is the most cause of colloid goiter.

6.065 If a patient's hypothalamus is destroyed, likely effects would include all of the following EXCEPT:
- A. a decreased responsiveness to the vasoconstrictor action of norepinephrine
- B. an increased secretion of prolactin
- C. a decreased secretion of estrogen
- D. a decreased basal metabolic rate
- E. an increased secretion GnRH

E. is correct
No increase in gonadotropin releasing hormone would be observed in the destruction of the hypothalamus.

6.066 Chronic hypersecretion of adrenal glucocorticoids will result in all of the following EXCEPT:
- A. increased degradation of muscle protein
- B. decreased concentration of ACTH in the blood
- C. hyperglycemia
- D. increased concentration of insulin in the blood
- E. improved wound healing

E. is correct
Glucocorticoid excess results in poor wound healing.

6.067 If the adipocyte membrane were freely permeable to glucose during fasting there would be:
- A. increased ketonemia
- B. increased mobilization of free fatty acids
- C. increased amounts of muscle glycogen
- D. hypoglycemia
- E. decreased gluconeogenesis

D. is correct
Glucose would freely enter the adipocyte resulting in hypoglycemia.

6.068 In humans, obesity has not been postively correlated with:
- A. type I diabetes
- B. increased coronary artery disease
- C. increased risk of hypertension
- D. elevated plasma leptin levels
- E. a greater incidence of specific cancerous growths

A. is correct
Type I diabetes usually develops before age 40 and is not associated with obesity. Plasma insulin levels are low and pathological changes in the B cells is often observed at the time of diagnosis.

6.069 Repeated surges of luteinizing hormone (LH) and follicle-stimulating hormone (FSH) are NOT seen during the luteal phase of the normal menstrual cycle because:
- A. concentrations of estrogen are too low
- B. progesterone blocks the positive feedback effects of estrogen
- C. pituitary stores of LH and FSH are depleted
- D. hypothalamic stores of luteinizing hormone releasing factors (LHRH) are not yet replenished
- E. the ovary has had insufficient time to produce a ripened follicle

B. is correct
High levels of progesterone block the positive feedback of estradiol during the luteal phase.

6.070 Release of growth hormone:
- A. is stimulated by somatostatin
- B. is elevated about one hour before the usual time of awakening
- C. as inhibited by dopamine
- D. occurs in slow-wave sleep (Stages III and IV)
- E. increases with age

D. is correct
Growth hormone release is associated with slow wave sleep.

**Use the following to answer questions 71-73.**

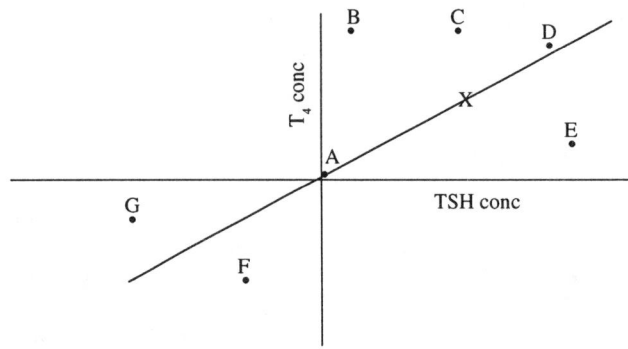

The line in the graph above represents the approximately normal relationship between plasma thyroxine ($T_4$) and thyroid-stimulating hormone (TSH). X is the normal value. Which point represents the steady-state effect of:

6.071  Hypophysectomy

A. is correct
There is no TSH and no $T_4$

6.072  Administration of thyroxine ($T_4$)

B. is correct
There is no TSH and increased $T_4$

6.073  Administration of propylthiouracil

E. is correct
Propylthiouracil blocks $T_4$ formation resulting in high TSH.

6.074  All of the following are affected by thyroid hormones EXCEPT:
   A. lipolysis
   B. differentation of the brain
   C. oxygen consumption by the adult brain
   D. basal metabolic rate
   E. heart rate

C. is correct
Oxygen consumption of the brain is normal in hypo- and hyperthyroidism.

6.075  Luteinizing hormone does not stimulate the release of adrenocortical hormones because:
   A. the major steroid output of the adrenal is corticosterone
   B. the adrenal gland does not possess cell membrane receptors specific for LH
   C. the adrenal gland does not possess an adenylate cyclase system
   D. LH does not contain the $\alpha$-subunit of ACTH
   E. ACTH counteracts the effects of LH

B. is correct
LH receptors are not present in the adrenal cortex.

6.076  A female patient develops a growth hormone-secreting tumor at age 32 and has elevated growth hormone levels for 10 years prior to detection of the tumor. All of the following would be present EXCEPT:
   A. increased stature
   B. bitemporal hemianopsia (visual field defects)
   C. protruding lower jaw (prognathism)
   D. enlarged liver
   E. lactation

A. is correct
At age 32, the patient has already finished her growth in stature with closure of epiphyses.

6.077 All of the following increase vasopressin secretion EXCEPT:
  A. water deprivation
  B. alcohol
  C. hypertonic saline
  D. 15% volume depletion
  E. nicotine

B. Ethanol is correct
Ehanol inhibits vasopressin release resulting in diuresis with alcohol ingestion.

6.078 Injection of insulin in a dose sufficient to produce hypoglycemia will also cause increased secretion of all of the following hormones EXCEPT:
  A. growth hormone
  B. pancreatic glucagon
  C. ACTH
  D. epinephrine
  E. thyroxine

E. is correct
Thyroxine secretion is not affected by insulin treatment. Thyroid hormones may increase the degraduation of insulin. The primary diabetogenic effect of thyroxine is to increase glucose absorption from the intestine.

6.079 All of the following are correct for hypothalamic and pituitary hormones EXCEPT:
  A. TRH is a potent releaser of prolactin
  B. SRIF will block release of STH
  C. prolactin is thought to be normally under tonic inhibitory control
  D. somatostatin is a tetradecapeptide
  E. dopamine is a primary stimulant of prolactin release

E. is correct
Dopamine is thought to be the hypothalamic prolactin-inhibiting hormone (PIH). Suckling of the breast removes the inhibitory influence of dopamine, resulting in prolactin release from the anterior pituitary.

6.080 Insulin administered to a diabetic patient produces all of the following effects EXCEPT:
  A. increased liver glycogen
  B. decreased urinary nitrogen excretion
  C. increased fat synthesis
  D. increased plasma free fatty acids
  E. increased respiratory quotient approaching unity

D. is correct
Insulin increases lipogenesis and does not have a lipolytic effect. A major manifestation of diabetes is increased lipid catabolism.

6.081 All the following are correct for iodide accumulation by thyroid cells EXCEPT:
  A. involves an active transport process
  B. is controlled in part by TSH
  C. is not directly affected by propylthiouracil
  D. is blocked by thiocyanate
  E. occurs at the plasma membrane-colloid interface

B. is correct
Iodide is transported from the circulation into the colloid. Iodide is pumped into the cell against an electrical gradient.

6.082 Glucagon differs from epinephrine in that it:
  A. does not activate phosphorylase in the liver
  B. does not promote glycogenolysis in the liver
  C. does not promote glycogenolysis in skeletal muscle
  D. does not elevate glucose significantly
  E. none of the above

C. is correct
Glucagon causes hepatic glycogenolysis but it does not cause glycogenolysis in muscle.

6.083  Which of the following increases both insulin secretion and glucagon secretion?
A. somatostatin
B. glucose
C. epinephrine
D. arginine
E. hypoglycemia

D. is correct
Amino acids (particularly glucogenic amino acids) are potent stimulators of both insulin and glucagon secretion.

6.084  The MSH activity in human plasma is mostly due to MSH in:
A. β-LPH
B. ACTH
C. α-MSH
D. β-MSH
E. β-endorphin

B. is correct
(alpha) MSH makes up amino acid residues 1-13 of the ACTH molecule. β-MSH is made up of the 17 amino acid residues at the C-terminal end of γ-LPH. Thus, ACTH has considerable MSH activity.

6.085  Which of the following produces the greatest increase in vasopressin release?
A. a 2% increase in plasma osmolality
B. a 2% decrease in blood volume
C. a 2% decrease in blood pressure
D. drinking a liter of water
E. drinking a liter of isotonic saline

A. is correct
When the effective osmotic pressure of plasma is increased by a little as 1%, vasopressin secretion is increased.

6.086  Which of the following is indicated by uterine bleeding following estrogen and progesterone withdrawal:
A. a normal endometrium
B. normal ovaries
C. normal pituitary
D. normal hypothalamus
E. normal adrenal cortex

A. is correct
Estrogen and progesterone are necessary for the growth and development of endometrium. Withdrawal of these hormones results in necrosis and sloughing off of the functional layer of the endometrium.

6.087  All of the following statements are correct for the action of insulin EXCEPT:
A. insulin activates hepatic glucokinase and glycogen synthetase
B. insulin promotes uptake of amino acids into muscle and stimulates protein synthesis
C. insulin reduces glucagon release, preserving energy stores
D. insulin activates both lipoprotein lipase and intracellular lipase of adipose tissue
E. insulin promotes transfer of glucose into adipose cells to provide alpha-glycero phosphate for re-esterification of fatty acids

D. is correct
Insulin stimulates lipogenesis and does activate lipoprotein lipase but inhibits intracellular hormone-sensitive lipase.

6.088  All of the following are true of oxytocin EXCEPT:
A. is thought to be primarily produced in the paraventricular nucleus of the hypothalamus
B. release can be elicited by certain emotional states and inhibited by alcohol
C. release is believed to initiate parturition
D. release can be described as a neuro-endocrine reflex
E. release in the female has been suggested to aid sperm transport

C. is correct
Oxytocin is not associated with the initiation of parturition and oxytocin levels are not increased in early labor.

6.089  Which of the following is correct concerning STH?
   A.  secretion can normally be suppressed by sufficient glucose administration
   B.  normally causes increased somatostatin production and release
   C.  normally stimulates lipogenesis in fat cells of the body
   D.  normally stimulates insulin receptor production and subsequent increased glucose uptake by cells of the body
   E.  exists in human plasma in at least two different forms.

A. is correct
Glucose infusions lower plasma growth hormone levels. Diabetic animals do not grow.

6.090  Ingesting a quart of milk would probably result in:
   A.  an increase in parathyroid hormone secretion
   B.  a reduction in the formation of 24, 25-dihydroxyvitamin $D_3$ in the kidney
   C.  a reduction in the formation of 1,25-dihydroxyvitamin $D_3$ in the kidney
   D.  tetany and hallucinations
   E.  a decrease in calcitonin secretion

C. is correct
Increased intestinal calcium absorption resulting from drinking a quart of milk will decrease the production of the active form of vitamin D and PTH.

**Use the following to answer questions 91-95**

   A.  adrenocorticotropic hormone    E.  prolactin
   B.  thyroid stimulating hormone    F.  somatomedin
   C.  follicle stimulating hormone    G.  oxytocin
   D.  luteinizing hormone    H.  vasopressin

6.091  Stimulates production of androgen binding protein

C. is correct
FSH stimulates the Sertoli cell to produce androgen binding protein (ABP).

6.092  Stimulates glucocorticoid secretion

A. is correct
The major function of ACTH is to stimulate cortisol synthesis. Cortisol is the principal glucocorticoid of the body.

6.093  Bromocriptine blocks secretion

E. is correct
Bromocriptine is a dopamine receptor agonist that inhibits prolactin release. Prolactin is under the inhibitory control of dopamine (prolactin inhibitory factor).

6.094  Propothiouracil administration increases release

B. is correct
Thiourea compounds block the iodination of tyrosine and thiouracil inhibits the conversion of $T_4$ to $T_3$. Thus, these effects increase TSH release.

6.095  Facilitates uterine contractions

G. is correct
Oxytocin induces contractions of the smooth muscle of the uterus.

**Use the following to answer questions 96-100.**

A. aldosterone      G. progesterone
B. epinephrine      H. testosterone
C. estradiol         I. thyroxine
D. estriol           J. oxytocin
E. cortisol          K. vasopressin
F. prolactin

6.096  Promotes sodium retention and potassium excretion

A. is correct
Aldosterone is the physiological mineralocorticoid of the body promoting sodium retention and potassium excretion.

6.097  Responsible for "milk let-down" during lactation

J. is correct
Oxytocin causes the contraction of myoepithelial cells resulting in milk ejection into the mammary gland ducts.

6.098  Secretion of the adrenal medulla

B. is correct
Epinephrine is the principal catecholamine released into the circulation when the adrenal medulla is stimulated.

6.099  Used as a marker of fetal well-being during high risk pregnancy

D. is correct
Plasma estriol levels are used as a marker for fetal well-being because estriol synthesis is dependent upon a functioning fetal adrenal and liver.

6.100  Secreted by the interstitial cells of Leydig

H. is correct
Testosterone is produced by the interstitial cells of Leydig of the testis.

# SECTION 7:   NEUROPHYSIOLOGY

7.001 The structure within the central nervous system whose function is the most rapidly affected by ischemia and anoxia is:
- A. neurilemma
- B. nerve cells
- C. astroglia
- D. microglia
- E. oligodendroglia

B. is correct
Neurons are the most sensitive of the structures listed to ischemia and anoxia.

7.002 The first sharp sensations of pain are carried by:
- A. type A-delta fibers
- B. type B fibers
- C. type C fibers
- D. type A-alpha fibers
- E. type D fibers

A. is correct
Pain impulses are transmitted by two fiber systems: small myelinated A-delta fibers and unmyelinated C fibers. A-delta fibers transmit their impulses to the CNS much faster than the C fibers.

7.003 Part of the functional role of the corpus callosum is thought to be related to:
- A. serving as another of several pathways for bilateral acoustic representation
- B. bilateral extraocular muscular synergism
- C. impressing memory engrams bilaterally
- D. distributing thalamic radiations to the cerebral cortex
- E. conveying part of the motor activity from one hemisphere to another

C. is correct
In split brain animals where the corpus callosum is sectioned no transfer of learning occurs.

7.004 All of the following sensory inputs are used to regulate normal posture EXCEPT:
- A. visual system
- B. vestibulae system
- C. muscle and joint receptors in the limbs
- D. neck muscle and ligament receptors
- E. auditory system

E. is correct
Sensory input from the auditory system is not directly used to regulate normal posture.

7.005 Auditory pathway conduction:
- A. can never occur from the olivary nuclei to the organ of Corti
- B. is inhibited by the sense of smell
- C. is not somatotopic
- D. shows a high preponderance of transmission in the contralateral pathway
- E. is modulated by efferent axons inhibit the organ of Corti's sensitivity

E. is correct
The auditory nerve contains efferent fibers originating near the superior olivary nucleus which decrease the sensitivity of the organ of Corti.

7.006 A specific thalamic nucleus that projects information associated with hearing is the:
- A. lateral geniculate body
- B. anterior nucleus
- C. ventroposterior lateral nucleus
- D. medial geniculate body
- E. ventroposterior medial nucleus

D. is correct
Auditory information is conveyed from the inferior colliculus to the medial geniculate body in the thalamus via the brachium of the inferior colliculus. The medial geniculate body projects to the primary auditory cortex.

7.007  In dementia due to Pick's disease, atrophic changes are most pronounced in the:
- A.  occipital cortex
- B.  basal ganglia
- C.  frontal and temporal lobes
- D.  cerebellar cortex
- E.  anterior horn of spinal cord

C. is correct

Pick's disease is a somewhat uncommon degenerative disorder in which neuron loss is primarily in the frontal and temporal cortex.

7.008  At operation, a surgeon exposes the postcentral gyrus and then stimulates this area. When the patient is asked to describe what he feels he would reply that:
- A.  he feels something is touching his index fingers
- B.  he feels heat in his foot
- C.  he perceives a bright light
- D.  he feels a tingling in his head
- E.  he has a bad headache

A. is correct

Sensations to the hand and finger are felt when this area is stimulated.

7.009  The neural substrate responsible for spontaneous involuntary movements is the:
- A.  cerebellum
- B.  basal ganglia
- C.  cerebral cortex
- D.  thalamus
- E.  hypothalamus

B. is correct

Involuntary movement disorders originate in one of the several basal ganglia.

7.010  The rate of formation of cerebrospinal fluid (CSF):
- A.  ensures total replacement of the contents of the ventricular system several times per day
- B.  increases with increased ventricular pressure and decreases with decreased ventricular pressure
- C.  normally exceeds the rate of CSF absorption
- D.  is abnormally increased in most patients with hydrocephalus

A. is correct

The volume of CSF is about 150 ml and the rate of production is about 550 ml/day. CSF replacement, therefore, occurs about 4 times per day.

7.011  Impulse conduction in axons can be blocked by:
- A.  pressure
- B.  cold
- C.  tetrodotoxin
- D.  ischemia
- E.  all of the above

E. is correct

All of these interfere with the opening or closing of voltage-gated sodium channels or maintenance of ionic gradients across the neural membrane.

7.012  Hypothalamic control of anterior pituitary function is maintained by:
- A.  direct vascular connections between the capillary beds of the corsal hypothalamus and the anterior pituitary
- B.  special neural projections from the median eminence of the hypothalamus to the anterior pituitary
- C.  hypothalamic secretion of hormones that influence the release of anterior pituitary hormones
- D.  initial formation of anterior pituitary hormones in the hypothalamus
- E.  direct connection between the strid terminalis and the anterior pituitary

C. is correct

Anterior pituitary secretion is controlled by chemical agents (hypophysiotropic hormones) that are carried in the portal hypophysial vessels from the hypothalamus to the pituitary.

**Use the following graph to answer question 13.**

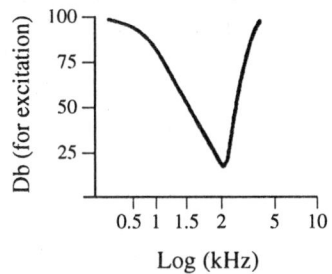

7.013  The auditory nerve fiber whose tuning curve is shown above:
    A.  cannot participate in pitch discrimination
    B.  has a threshold of less than 25 decibels
    C.  innervates hair cells nearest the base of the cochlea
    D.  responds only to sound having a frequency of 2 kHz.
    E.  will not respond to sounds over 100 decibels

B. is correct
Threshold is the lowest decibel level at which an auditory nerve fiber responds. Threshold for this nerve fiber is at 2 kHz.

**Use the following to answer questions 14 and 15:**

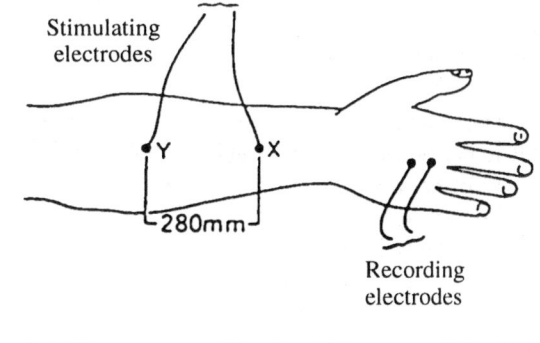

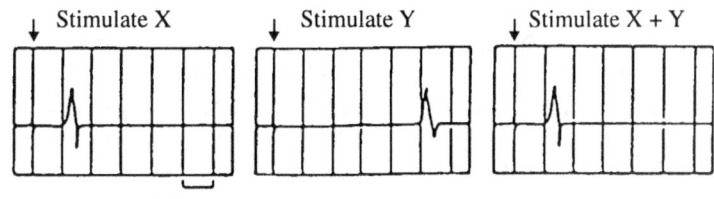

1.0 msec

7.014  Electromyograms were made from a muscle in a man's hand. The motor nerve was stimulated through either of two electrodes (X and Y) separated by 280 mm on the arm. The conduction velocity of the motor axons (in m/sec) is approximately:
    A.  30
    B.  50
    C.  70
    D.  90
    E.  110

C. is correct
The difference in the conduction time between stimulating at X and Y is about 40 msec; then 280 mm ÷ 40 msec is 70 m/s.

7.015  When stimuli at X and Y are applied
simultaneously, only one response is seen because the:
   A. antidromic impulses generated at X collide with
      orthodromic impulses generated from Y
   B. impulses initiated at Y are slowed by partial
      refractoriness at X
   C. muscle is refractory to impulses arriving from Y
   D. nerve at X is refractory to the impulses arriving
      from Y
   E. responses to stimuli at X and Y occur
      simultaneously

A. is correct
Stimuli at X and Y initiate nerve impulses in
both the orthodromic and antidromic direction.
When the orthodromic impulse generated from Y
interacts with the absolute refractory period of
the antidromic impulse generated at X, the
orthodronic impulse from Y is blocked.

7.016. All of the following responses occur during
accommodation for near vision EXCEPT:
   A. increased curvature of the lens
   B. relaxation of the ciliary muscle
   C. contraction of the medial rectus muscle
   D. contraction of the radial muscle of the iris
   E. increased refractive power of the eye

B. is correct
When looking at a near object, the ciliary muscle
contracts to relax tension on the ciliary fibers
allowing the lens to round up.

7.017  In comparison with cones, rods:
   A. are more common in the central retina
   B. are more involved in color vision
   C. yield a greater acuity of vision
   D. have a lower threshold to light
   E. have their maximum sensitivity in bright light

D. is correct
The rods are specialized for vision in low levels
of light. Thus, visual activity is sacrificed for
dark adaptation.

7.018  Bitemporal hemianopsia visual field defects can
be associated with lesions of the:
   A. visual cortex
   B. lateral geniculate body
   C. optic radiation
   D. optic (II) nerve
   E. optic chiasm

E. is correct
Lesions of the optic chiasm sever the crossing
optic nerve fibers which carry visual information
from both lateral visual fields. Such a lesion can
be caused by a pituitary tumor pushing on the
chiasm.

7.019  Color vision is mediated by three different
photopigments:
   A. all three of which are present in rods only
   B. all three of which are present in cones only
   C. all three of which are present in both rods and cones
   D. two of which are present in rods and one of
      which is present in cones
   E. one of which is present in rods and two of which
      are present in cones

B. is correct
It is postulated that color vision is due to three
kinds of cones, each sensitive to one of the three
primary colors with the sensation of any given
color determined by the frequency of impulses
from each of these cone systems.

7.020  Instant blindness in one eye would occur from
transection or destruction of the:
   A. optic tract
   B. optic nerve
   C. right half of the optic chiasm
   D. postcentral gyrus
   E. temporal lobe

B. is correct
Destruction or transection of one optic nerve will
result in blindness of that eye, whereas
transection of one optic tract results in blindness
in half of the visual field.

7.021 All of the following statements about spinal nerves are true EXCEPT:
- A. all spinal nerves are formed by the union of dorsal and ventral rootlets
- B. in the thoracic region, dorsal and ventral rami do not join with other rami
- C. dorsal rootlets transmit only sensory information
- D. each spinal nerve divides into a dorsal and ventral ramus
- E. both dorsal and ventral rami contribute to the formation of the cervical plexus

E. is correct
Only the ventral rami join to form the cervical, brachial, and lumbosacral plexi.

7.022 All of the following statements concerning the ear are true EXCEPT:
- A. the semicircular canals, vestibule and cochlea are a part of the osseous labyrinth
- B. the membranous labyrinth is filled with perilymph
- C. the anterior labyrinth of the inner ear (cochlea) is involved with hearing
- D. the cochlea consists of a central modiolus with a spiral lamina extending into the lumen of the cochlea
- E. the spiral lamina houses bipolar neuron cell bodies of the 8th cranial nerve

B. is correct
The membranous labyrinth is filled with endolymph, having ionic consistency similar to intracellular fluid.

7.023 Acting as a general arousal and attention system is an important function of the:
- A. hypothalamus
- B. occipital cortex
- C. reticular formation
- D. colliculi
- E. basal ganglia

C. is correct
The reticular formation is located in the tegmentum of the entire brainstem. Its projections to the driencephalon and telencephalon, including cholinergic, noradrenergic and serotonergic fibers, constitute the ascending reticular activating system. It also participates in producing sleep cycles.

7.024 The basal ganglia include all of the following EXCEPT:
- A. globus pallidus
- B. amygdaloid nucleus
- C. hippocampus
- D. putamen
- E. caudate nucleus

C. is correct
The hippocampus is part of the limbic system and is not part of the basal ganglia.

7.025 With respect to the meninges:
- A. the bulk of the cerebrospinal fluid is present between the dura and arachnoid
- B. a relatively wide cleft is present between the dura covering the brain and the endosteum of the skull
- C. the pia mater is intimately adherent to the surface of the brain
- D. in the cranial cavity the dura is fused with the pia
- E. in the cranial cavity extensions from the pia mater form septa between various parts of the brain

C. is correct
The meninges and CSF protect the brain. The arachnoid which is fused to the dura by the surface tension of the thin layer of fluid between the two membranes, projects web-like trabeculae which attach to the pia mater and form the subarachnoid space.

7.026  In executing a movement bringing the extended arm and index fingers pointing and aiming toward a given target, at which the finger has to stop, the cerebellum is responsible for the:
  A. decreased tone of the agonist muscle
  B. increased tone of the extensors
  C. the tremor associated with such movements
  D. damping of muscle contraction necessary to avoid overshooting the target
  E. increased tone of the agonist muscles

D. is correct
The cerebellum smoothes and coordinates movements. A cerebellar hemisphere lesion results in overshooting when attempting to touch an object with the finger.

7.027  The term echopraxia refers to the:
  A. meaningless repetition of words
  B. ability to reproduce echoes
  C. inability to distinguish shapes by feel
  D. subjective experience of hearing echoes
  E. repetition of observed movements

E. is correct
Echopraxia is the automatic initation of observed movements.

7.028  Sherrington described the anterior horn cell as "the final common pathway", with nothing occurring to influence neuronal transmission in the central nervous system after this point. A generalized CNS stimulant which acts by inhibiting post-synaptic inhibitory mechanisms mainly in the spinal cord is:
  A. picrotoxin
  B. dextroamphetamine
  C. pentylenetetrazol
  D. strychnine
  E. doxapram

D. is correct
Strychnine blocks inhibitory receptors for glycine. These are mainly located in the spinal cord.

7.029  All of the following statements regarding cholinergic blockade are true EXCEPT:
  A. muscarinic actions of all parasympathomimetic drugs are blocked selectively by atropine, through competitive occupation of cholinergic receptors
  B. epinephrine and other sympathomimetic amines antagonize most muscarinic effects at sites where adrenergic and cholinergic impulses produce opposing effects
  C. nicotinic effects of acetylcholine and its derivatives at autonomic ganglia are blocked by hexamethonium
  D. nicotinic effects of acetylcholine and its derivatives at the neuromuscular junction of skeletal muscle are antagonized by D-tubocurarine
  E. muscarinic effects of acetylcholine and its derivatives at autonomic ganglia are blocked by hexamethonium

E. is correct
Muscarinic effects of acetylcholine are blocked by atropine or scopolamine which block muscarinic receptors.

7.030  Contraction of the urinary bladder results from stimulation of:
  A.  sympathetic innervation
  B.  parasympathetic innervation
  C.  pudendal nerves
  D.  visceral afferent pathway
  E.  none of the above

B. is correct
Contraction of the smooth muscle of the bladder know as the detrusor muscle is mainly responsible for emptying the bladder. Parasympathetic fibers in the pelvic nerves initiate a reflex contraction.

7.031  In most reflex arcs impulses are transmitted from:
  A.  interneuron to afferent neuron
  B.  efferent neuron to interneuron
  C.  interneuron to efferent neuron
  D.  efferent neuron to afferent neuron
  E.  afferent neuron to effector cell

C. is correct
Many reflexes have interneurons between the afferent and efferent neurons.

7.032  Group I primary afferent axons (type Ia) that innervate muscle spindles or tendon organs (type Ib) can be differentiated most reliably on the basis of:
  A.  axonal conduction velocities
  B.  the muscle in which they appear
  C.  thresholds to electrical stimulation
  D.  responses to muscle contraction
  E.  regularity of static discharge

D. is correct
The most reliable test for type Ia afferents is the decrease or cessation of their action potentials following a muscle contraction.

7.033  All of the following statements concerning the pars distalis of the pituitary are correct EXCEPT:
  A.  its blood supply is the hypophyseal portal vessels
  B.  its nerve supply is the hypothalamo-hypophyseal tract
  C.  it is separated from the pars nervosa by the pars intermedia
  D.  one of the basophils secretes follicle stimulating hormone (FSH)
  E.  it has acidophils which secrete somatotropin (growth hormone)

B. is correct
The pars distalis does not have an innervation from the hypothalamus. It receives neurohumoral control from the hypothalamus via the hypophyseal portal vessels.

7.034  In the adult the spinal cord ends inferiorly at which vertebral level?
  A.  T-12
  B.  L-2
  C.  L-5
  D.  the sacrum
  E.  S-5

B. is correct
The spinal cord tissue ends at about the L2-3 interspace in adults. In neonatal children it ends at about L-5.

7.035  The large number of lumbosacral roots surrounding the filum terminale is known as the:
  A.  coccygeal ligament
  B.  denticulate ligament
  C.  conus medullaris
  D.  cauda equina
  E.  lumbar cistern

D. is correct
The cauda equina is composed of dorsal and ventral roots emanating from the spinal cord and traversing the intravertebral space below L-2 to reach their respective level of exit.

7.036 Compared to large motor neurons, small motor neurons are more:
- A. depolarized by the same excitatory current
- B. capable of discharging without synaptic input
- C. likely to innervate glycolytic muscle fibers
- D. likely to be silent during maintained postures
- E. likely to be silent during maximal contractions

A. is correct
The larger input resistance of the smaller motor neurons allows them to depolarize more for the same excitatory current than do large motor neurons.

7.037 In humans, complete transection of spinal cord at the level of T-2 leads to each of the following EXCEPT:
- A. abolition of the penile erection reflex
- B. establishment of a spinal voiding reflex after 1-5 weeks
- C. loss of conscious sensations below T-2
- D. loss of voluntary motor activity below T-2
- E. presence of an extensor plantar response after the patient has recovered from spinal shock

A. is correct
Coordinated sexual activity is absent after spinal cord transection. Penile erection, however, remains possible with genital stimulation.

7.038 During an episode of severe stress, the entire sympathetic nervous system may become active and discharge neurotransmitter. Which of the following is NOT a direct result of sympathetic discharge?
- A. blanching of the skin
- B. cold sweat
- C. dilated pupils
- D. involuntary defecation
- E. rapid heart rate

D. is correct
The defecation reflex is under the control of the parasympathetic nervous system. Discharge of the sympathetic nervous system is associated with inhibition of motor activity in the gut.

7.039 The major neurotransmitter released by cerebellar Purkinje cells is:
- A. $\gamma$-aminobutyric acid (GABA)
- B. aspartate
- C. glutamate
- D. norepinephrine
- E. serotonin

A. is correct
The output of the cerebellum, Purkinje cell axons, is inhibitory via the neuron transmitter GABA.

7.040 Lesions associated with movement disorders (e. g., chorea and athetosis) are usually:
- A. related to $\gamma$-motor neuron control via the reticular formation
- B. related to spinal recurrent inhibition of motoneurons
- C. within the basal ganglia
- D. within the cerebellum
- E. within the primary motor cortex

C. is correct
Basal ganglia circuits regulate initiation, speed and amplitude of movements. Disease of the basal ganglia cause abnormalities of movement known as involuntary movement disorders.

7.041 The most prominent cluster of noradrenergic neurons in the central nervous system is found in the:
- A. locus coeruleus
- B. medial raphe nucleus
- C. nucleus basalis
- D. parvicellular division of the substantia nigra
- E. Purkinje cells of the cerebellum

A. is correct
The cell bodies of the noradrenaline containing neurons in the brain are located in the locus coeruleus.

7.042  All of the following sensory systems relay in the thalamus before going to the cortex EXCEPT:
  A.  olfactory system
  B.  visual system
  C.  auditory system
  D.  gustatory system
  E.  somatosensory system

A. is correct
All sensory systems project first to the thalamus before going on to the cerebral cortex except for the olfactory system.

7.043  All neurons with cell bodies in the spinal cord and axon terminals outside the central nervous system:
  A.  are excitatory-adrenergic in function
  B.  are excitatory-cholinergic in function
  C.  are unmyelinated
  D.  exert effector control through electrically operating synapses
  E.  terminate on muscle or gland cells

B. is correct
Both types of neurons in the spinal cord which project axons into the peripheral neurons system, motor neurons and preganglionic antonomic neurons, use acetylcholinic as their transmitter.

7.044  Indifference toward a sensory deficit and neglect of stimuli is most likely to occur with lesions in the:
  A.  amygdala on either side
  B.  left inferior frontal gyrus (Broca's area)
  C.  left posterior superior temporal gyrus (Wernicke's area)
  D.  right hippocampus
  E.  right parietal association cortex

E. is correct
Lesions in the parietal association area, particularly in the posterior part, lead to unilateral inattention and neglect that causes patients to ignore stimuli from the contralateral portion of their bodies in the non-dominant hemisphere.

7.045  Unmyelinated (c) fibers of mammalian nerves:
  A.  are responsible for slow pain sensations
  B.  are blocked last by local anesthetics
  C.  provide efferent preganglionic sympathetic innervation to most viscera
  D.  are insensitive to pressure applied to the nerve
  E.  are not blocked by local anesthetics because they do not have nodes of Ranvier

A. is correct
Unmyelinated C fibers conduct pain at a low velocity.

7.046  Nerve endings that subserve the sensation of pain are usually:
  A.  in the form of laminations of fibers in the dermis
  B.  supplied by a large group of fibers from many neurons
  C.  bare endings of myelinated or unmyelinated fibers
  D.  different in structure in somatic and autonomic divisions of the nervous system
  E.  larger in the more sensitive areas like hands and face

C. is correct
Pain receptors are bare, unencapsulated nerve endings.

7.047  In humans, bilateral hippocampal lesions are associated with:
  A.  greatly increased aggressiveness
  B.  intellectual impairment
  C.  permanent anosmia
  D.  recent memory loss
  E.  visual field defects

D. is correct
Bilateral destruction of the hippocampus results in significant defects in recent memory, as in Korsakoff's syndrome.

7.048 Which of the following defects would result from a selective loss of cones in the retina?
- A. a decrease in critical fusion frequency
- B. diplopia
- C. enhanced lateral inhibition
- D. fovea blindness
- E. night blindness

D. is correct
The fovea centralis is composed of densely packed cones and no rods. Thus, a selective loss of cones would result in fovea blindness.

7.049 Radial muscle contraction results in:
- A. a decrease in the convexity of the crystalline lens
- B. an increase in the anteroposterior dimension of the eyeball
- C. an increase in the convexity of the crystalline lens
- D. pupillary constriction
- E. pupillary dilation

E. is correct
Contraction of this muscle causes pupillary dilation under sympathetic control.

7.050 In the visual system, the equivalent of cutaneous two-point threshold discrimination is:
- A. a shift of sensitivity during adaptation to darkness
- B. discrimination of brightness
- C. visual acuity
- D. discrimination of color
- E. flicker fusion

C. is correct
Visual acuity is a measure of the ability to detect two lines with a minimum separation distance as separate.

7.051 Ganglion cells of the retina are known to:
- A. Receive lateral inhibition mediated by horizontal and amacrine cells
- B. lack the capacity to determine a contrast between dark and light areas of vision
- C. increase their frequency of discharge at a constant rate
- D. transmit nerve impulses at rate of 55 per second, even in the dark
- E. respond to the suppression of light by decreasing their rate of discharge in all cases

A. is correct
Lateral inhibition, which enhances visual acuity, is a prominent feature shaping retinal ganglion cell responses.

7.052 Normal sleep is composed of two types of sleep, rapid eye movement sleep (REM sleep) and non-REM sleep. Non-REM sleep is further divided into stages 1, 2, 3, and 4. With increasing age one has:
- A. more REM and more stage 4 sleep
- B. more REM and less stage 4 sleep
- C. less REM and more stage 4 sleep
- D. about the same REM and less stage 4 sleep
- E. the amounts of REM and stage 4 do not change with age

D. is correct
With age slow wave sleep declines more than other types of sleep. By age 60 slow wave sleep may be reduced greatly, more so in men.

7.053 After selective deprivation of REM sleep, a person will show:
- A. a decrease in REM-associated penile tumescence
- B. an increase in arm and leg movements during sleep
- C. a later onset of REM sleep during subsequent sleep
- D. less intensive REM sleep (fewer phasic eye movements)
- E. the REM rebound phenomenon

E. is correct
Following deprivation of the REM sleep stage specifically, REM rebound occurs, in which the REM sleep stage becomes longer and more intense, having more eye movements and more intense dreams

7.054 All of the following statements concerning type Ia axons, the largest myelinated (A) fibers of mammalian somatic nerves are true EXCEPT:
  A. conduct at velocities greater than 100 m/sec
  B. convey information about muscle stretch to the central nervous system
  C. synapse directly on alpha motor neurons
  D. may lead to a withdrawal reflex when activated
  E. project to the dorsal nucleus of Clarke

D. is correct
The type Ia muscle spindle afferents have the largest diameter of any axon in the mammalian neurons system, and hence, the fastest conduction velocities. They convey information for the monosynaptic reflex and to cells of the dorsal spinocerebellar tract; the dorsal nucleus of Clarke.

7.055 Receptors responding to changes of the osmolality of body fluids can be found in the:
  A. lateral hypothalamus
  B. carotid bodies
  C. supraoptic nucleus of the hypothalamus
  D. olfactory cells
  E. paraventricular nucleus of the hypothalamus

C. is correct
An increase in the osmotic pressure of the blood perfusing the supraoptic nucleus cells increases the release of antidiuretic hormone from these cells.

7.056 In a normal awake subject who is supine with his head elevated by 30 degrees, irrigation of the left external auditory canal with cold water will cause:
  A. nystagmus with the fast phase to the left
  B. nystagmus with the fast phase to the right
  C. tonic deviation of the eyes to the left
  D. tonic deviation of the eyes to the right

B. is correct
Eye movements following caloric stimulation are described by the mnemonic COWS (cold-opposite; warm-same) where the direction is that of the fast phase of nystagmus.

7.057 Following section of the nerve supply to skeletal muscle, the:
  A. basal lamina at the end-plate region appears
  B. muscle becomes supersensitive to application of acetylcholine
  C. muscle fibers become electrically inexcitable
  D. presynaptic terminal contains greater numbers of synaptic vesicles than before the nerve section
  E. Schwann cell shealth degenerates

B. is correct
After denervation, the muscle fiber makes a less mature form of acetylcholinic receptor which is then found widespread over the muscle cell membrane, not just at the neuromuscular junction, making it more sensitive to ACh. This form of the acetylcholine receptor is normally present during the development and innervation of muscle fibers.

7.058 Which of the following physiologic characteristics is common to receptor potentials, and excitatory and inhibitory postsynaptic potentials?
  A. they are graded potentials
  B. an increase in permeability to potassium ions
  C. an increase in permeability to sodium ions
  D. a refractory period
  E. a hyperpolarizing phase

A. is correct
These potentials vary in size according to the level of stimulus in the case of receptor potentials or the degree of convergence and summation in the case of excitatory and inhibitory postsynaptic potentials.

7.059 A 50-year-old man does outdoor construction work in a very hot and humid climate. An increase in which of the following mechanisms increases his rate of heat loss?
  A. parasympathetic stimulation of the sino-atrial node of the heart
  B. somatic stimulation of skeletal muscles
  C. sympathetic adrenergic stimulation of the arteriovenous anastomoses of the skin
  D. sympathetic adrenergic stimulation of peripheral arterioles
  E. sympathetic cholinergic stimulation of sweat glands

E. is correct
Stimulation of sympathetic cholinergic impulses to sweat glands results in an increased generalized secretion of sweat. Noradrenergic impulses result in a slight localized secretion.

7.060  In fever caused by bacterial infection, endogenous pyrogens in the blood:
  A. act upon skeletal muscle to increase metabolic activity
  B. directly inhibit secretory activity of the sweat glands
  C. directly induce both shivering and vasodilation
  D. act to reset the hypothalamic "thermostat" to regulate at a higher temperature
  E. directly attack and destroy infectious disease organisms

D. is correct
Endogenous pyrogens produced during infection appear to act directly on thermoregulatory centers.

7.061  During a fever, the rise in core temperature:
  A. is reduced by agents that inhibit synthesis of prostaglandins
  B. involves the formation of endogenous pyrogens
  C. results from an increase in heat production and a decrease in heat loss
  D. involves the same physiologic mechanisms as the temperature rise during exercise
  E. is proportional to the amount of brown fat present

A. is correct
Agents that inhibit prostaglandin synthesis (e.g. aspirin) act on the hypothalamus to inhibit prostaglandin synthesis.

7.062  Bilateral lesions in the posterior hypothalamus result in:
  A. increased activity in the sympathetic cholinergic innervation of the sweat glands
  B. activation of hypothalamic thermoreceptors
  C. a condition in which body temperature varies with the environment
  D. decreased activity in the sympathetic adrenergic innervation mediating vasoconstriction in the skin
  E. horripilation of hair on the skin

C. is correct
Bilateral lesion in the posterior hypothalamus result in poikilothermia, a condition in which body temperature rises and lowers with the environment.

7.063  A lesion in the facial area in the motor strip of the precentral gyrus most likely will result in:
  A. facial apraxia
  B. expressive aphasia
  C. facial paresis
  D. facial paresthesia
  E. receptive aphasia

C. is correct
This is an upper motor neuron lesion which renders the facial muscles weak or paralyzed in voluntary movements but active in involuntary movements

7.064  Deafness in one ear would be produced by destruction of which structure on the ipsilateral side?
  A. auditory cortex
  B. cochlear nucleus
  C. inferior colliculus
  D. lateral lemniscus
  E. medical geniculate body

B. is correct
Destruction of the cochlear nucleus blocks auditory information from one ear from reaching the central auditory pathway. In the other structures, auditory information is bilaterally derived.

7.065  In vertebrate skeletal muscle, contraction is triggered when calcium binds to:
  A. F-actin
  B. heavy meromyosin
  C. myosin light-chain kinase
  D. tropomyosin
  E. troponin

E. is correct
The electrical impulse causes release of $Ca^{++}$ from the endoplasmic reticulum which then binds to troponin.

7.066 Which of the following explains the steady contractile force of skeletal muscle in rigor mortis?
  A. myosin cross-bridges are rapidly autolyzed
  B. the concentration of ATP falls below that required for relaxation
  C. the concentration of calcium is too low for relaxation
  D. T tubules are disrupted
  E. tropomyosin is no longer activated by calcium

B. is correct
When skeletal muscle fibers are depleted of ATP and phosphorylcreatine, they develop a state of extreme rigidity called rigor. After death it is called rigor mortis.

7.067 A 4-year-old boy becomes febrile, lethargic, and loses weight. Physical examination reveals a mass in the region of the upper pole of the left kidney. Surgical exploration reveals a tumor measuring 10 cm in diameter, which is excised. Microscopic examination discloses small round cells with pseudo-rosette formation. The tumor is a(n):
  A. neuroblastoma
  B. pheochromocytoma
  C. carcinoma
  D. adenoma
  E. ganglioneuroma

B. is correct
The clinical manifestations are associated with increased secretion of catecholamines. Increased metabolic rate may cause weight loss, impaired heat loss and feelings of fatigue or exhaustion.

7.068 Activation of γ-motoneurons can induce contraction of skeletal muscle by:
  A. direct innervation of extrafusal muscle fibers
  B. inhibition of muscle spindle receptors
  C. monosynaptic connections with α-motoneurons
  D. reflex activation of Golgi tendon organs
  E. reflex activation through contraction of intrafusal muscle fibers

E. is correct
The gamma motoneurons are small neurons in the ventral horn of the spinal cord that innervate the intrafusal fibers of the muscle spindles. Contraction of intrafusal muscle fibers of the muscle spindles activate type Ia afferents which monosynaptically activate alpha motor neurons innervating the same muscle.

7.069 Anterograde amnesia is usually associated with a lesion in the:
  A. frontal lobe
  B. thalamus
  C. midbrain
  D. parietal lobe
  E. limbic system

E. is correct
Anterograde amnesia refers to the inability to transfer new information from short term to long term memory. This loss of recent memory is seen in alcohol abuse and correlates with damage to the hippocampus due to thiamine deficiency.

7.070 Each pharyngeal arch has its own specific cranial nerve which innervates the structures derived from that arch. The nerve of the second arch is the:
  A. facial
  B. hypoglossal
  C. pharyngeal
  D. trigeminal
  E. vagus

A. is correct
The facial nerve innervates muscles derived from the second pharyngeal (branchial) arch.

**Use the following to answer questions 71-74.**

    A. lateral corticospinal tract
    B. lateral reticulospinal tract
    C. rubrospinal tract
    D. anterior corticospinal tract
    E. vestibulospinal tract

For each function or projection of a descending pathway, select the correct tract.

7.071 A completely crossed tract projecting largely to distal flexor interneurons no direct connections to motoneurons.

C. is correct
The rubrospinal tract is similar to the corticospinal tract in termination but mainly influences the upper limbs and has no monosynaptic synapses on motor neurons.

7.072 This tract causes inhibition of extensor and flexor muscles.

B. is correct
The lateral reticulospinal tract appears to mediate the inhibition associated with REM sleep.

7.073 This tract terminates directly on motoneurons (almost exclusively distal flexors) and on interneurons that influence motoneurons.

A. is correct
A hallmark of the corticospinal tract is fine motor control of the hands.

7.074 Unilateral destruction drastically reduces decerebrate rigidity on the same side of the lesion.

E. is correct
The vestibulospinal tract strongly excites extensor motor neurons on the ipsilateral side.

**Use the following to answer questions 75 and 76.**

    A. cerebellar cortex
    B. sensory cortex
    C. red nucleus
    D. substantia nigra pars compacta
    E. pre-motor cortex (Brodmann area 6)

For each lesion, select the correct structure.

7.075 Lesion can lead to difficulty in initiating movement.

D. is correct
Loss of substantia nigra dopaminergic cells is a hallmark of Parkinson's disease in which difficulty in initiating movements is one symptom.

7.076 Lesion can lead to spasticity with an increased stretch reflex.

E. is correct
Lesion of the premotor cortex produces the upper motor neuron syndrome in which hyperactive stretch reflexes are evident.

**Use the following to answer questions 77 and 78.**

    A. myopia
    B. presbyopia
    C. astigmatism
    D. hyperopia
    E. anisocoria

For each condition below select the correct name from the list above.

7.077  Correctable by diverging lenses

A. is correct
Myopia can be corrected by divergent lens with minus diopters.

7.078  Anteroposterior length of the eyeball is too short for an unaccommodated lens system

D. is correct
In hyperopia, the eyeball is relatively too short and the image is focused behind the retina.

**Use the following to answer questions 79-81.**

    A. Golgi tendon organ
    B. pacinian corpuscle
    C. Meissner's corpuscle
    D. free nerve ending
    E. hair follicle

For each statement below, select the correct receptor from the list above.

7.079  Receptor which responds to both stretch and contraction of a muscle

A. is correct
The Golgi tendon organ being in series with the muscle tendon responds to tension in the tendon from either stretch or contraction.

7.080  Responds to vibratory stimuli

B. is correct
Pacinian corpuscles transduce vibratory stimuli as part of proprioception.

7.081  An encapsulated receptor for touch

C. is correct
Meissner's corpuscles are found in glabrous skin (hairless) and are sensitive detectors of touch.

**Use the following to answer questions 82 and 83.**

    A. cerebral cortex
    B. basal ganglia
    C. thalamus
    D. hypothalamus
    E. cerebellum

For each lesion below, select the correct structure from the list above.

7.082 Lesions can result in a tendency toward poikilothermy

D. is correct
Lesions in the posterior hypothalamus causes the body temperature to fall in relation to the environment.

7.083 Lesions can result in development of athetoid movement

B. is correct
Lesions of the basal ganglia produce involuntary contractions in an otherwise inactive individual. Athetosis may be produced by a lesion in the lenticular nucleus composed of the caudate and putamen nuclei.

7.084 A lesion of which of the following hypothalamic nuclei results in obesity?
    A. supraoptic
    B. ventromedial
    C. dorsomedial
    D. suprachiasmatic
    E. lateral

B. is correct
Lesions of ventromedial hypothalamic nuclei result in obesity (Frolich's Syndrome).

**Use the following to answer questions 85 and 86.**

    A. ventroposterior lateral nucleus
    B. lateral geniculate
    C. ventroposterior medial nucleus
    D. dorsomedial nucleus
    E. centromedian nucleus

For each statement below, select the correct structure from the list above.

7.085 Projects somatic sensation from the face to the primary somatosensory cortex

C. is correct
The ventroposterior medial nucleus receives information from the trigeminal sensory nuclei and projects to the postcentral gyrus for conscious appreciation.

7.086 This diffusely projecting nucleus is part of the arousal system

E. is correct
The centromedian nucleus receives input from the reticular activating system and projects to wide areas of the cerebral cortex.

**Use the following to answer questions 87-90.**

A. acetylcholine          F. glutamate/aspartate
B. adenosine              G. glutamate
C. dopamine               H. histamine
D. epinephrine            I. norepinephrine
E. γ-aminobutyric acid (GABA)   J. serotonin

For each effect, select the appropriate neurotransmitter.

7.087 Released by raphe neurons projecting to the spinal cord for analgesic action

J. is correct
Serotonergic fiber systems have been found to inhibit transmission of pain sensation in the dorsal horn.

7.088 Produced by the neurons of the nucleus basalis of Meynert that degenerate in primary degenerative dementia, Alzheimer type

A. is correct
In Alzheimer's disease there is large loss of cholinergic neurons in the nucleus basalis of Meynert.

7.089 This unusual neurotransmitter is present in high concentration in the putamen and caudate nuclei

C. is correct
In Parkinsonism, dopaminergic neurons of the substantia nigra are destroyed and this leads to reduced output of the caudate and putamen nuclei.

7.090 An inhibitory neurotransmitter that increases the permeability of a neuron to chloride

E. is correct
GABA serves to stabilize the membrane potential at the equilibrium potential for Cl⁻ by increasing the permeability of the membrane to Cl⁻.

**Use the following to answer questions 91-93.**

A. vestibulocochlear nerve     G. optic nerve
B. facial nerve                H. spinal accessory nerve
C. glossopharyngeal nerve      I. trigeminal nerve
D. hypoglossal nerve           J. trochlear nerve
E. oculomotor nerve            K. vagus nerve
F. olfactory nerve

**Match the statements below with the cranial nerve from the list above.**

7.091 A patient comes to the hospital and reports difficulty in swallowing. His resting heart rate is 110 beats per minute.

K. is correct
Impaired vagal function will disrupt swallowing and allow the heart rate to increase.

7.092 For the past week a 56-year-old patient has complained of dizziness, vomiting, and nausea. She states that often the room appears to be spinning. In the examination room she has difficulty getting on and off the table.

A. is correct
Abnormal function of the vestibular portion of cranial n. VIII will produce vertigo, nausea, vomiting, and balance problems.

7.093  A 21-year old male is brought to the emergency room after a fall, complaining that his vision is failing. On further examination it is found that his left eye deviates down and laterally.

E. is correct
Impairment of the oculomotor nerve allows the eye to be deviated down and laterally by the two non-impaired extraocular muscles.

**Use the following to answer questions 94-100.**

A. sympathetic nervous system
B. parasympathetic nervous system
C. cerebellum
D. temporal lobe
E. occipital lobe
F. optic chiasm
G. optic tract
H. optic nerve
I. lower motor neuron
J. upper motor neuron
K. superior colliculus
L. inferior colliculus
M. acetylcholine
N. norephinephrine

**For each statement below select the correct structure from the list above.**

7.094  This type of lesion produces hyperreflexia and an extensor plantar response

J. is correct
Lesions of the corticospinal tract anywhere along its course produces the signs of hyperactive deep tendon reflex and the Babinski sign (extensor plantar response).

7.095  Neurotransmitter belonging to the sympathetic nervous system that innervates sweat glands

M. is correct
Sympathetic innervation to the sweat glands is cholinergic.

7.096  A lesion here produces a lack of response to a sudden peripheral visual cue

K. is correct
The superior colliculus is involved in moving the eyes reflexively in response to a visual cue appearing suddenly in the visual field.

7.097  Secreted secondary to vagal stimulation

M. is correct
Acetylcholine is the primary neurotransmitter for parasympathetic nerves.

7.098  A lesion here results in a loss of vision in the temporal visual fields bilaterally

F. is correct
Lesion of the optic chiasm produces a bitemporal hemianopsia visual feld loss.

7.099  The most common site of epileptiform discharge

D. is correct
Epilepsy is most commonly associated with limbic structures in the temporal lobe, the hippocampus and amygdala.

7.100  A lesion here results in a gait similar to that seen in alcohol intoxication

C. is correct
Lesions of the midline structure of the cerebellum (vermis) result in ataxia, similar to the gait seen in alcohol intoxication.